Zoophysiology Volume 14

A. D. Hasler · A. T. Scholz

Olfactory Imprinting and Homing in Salmon

Investigations into the Mechanism of the Imprinting Process

In Collaboration with R. W. Goy

With 25 Figures

Springer-Verlag
Berlin Heidelberg New York Tokyo 1983

Prof. ARTHUR D. HASLER
University of Wisconsin
Center for Limnologie
Madison, WI 53706, USA

Prof. ALLAN T. SCHOLZ
Eastern Washington University
Department of Biology
Cheney, WA 99004, USA

The cover illustration shows a morpholine molecule associated with outside nose nerve and thyroxine molecule associated with brain.

ISBN-13:978-3-642-82072-4 e-ISBN-13:978-3-642-82070-0
DOI: 10.1007/978-3-642-82070-0

Library of Congress Cataloging in Publication Data. Hasler, Arthur D. (Arthur Davis), 1908–. Olfactory imprinting and homing in salmon. (Zoophysiology; v. 14). Includes bibliographical references and index. 1. Salmon – Homing. 2. Imprinting (Psychology). 3. Odors. 4. Smell. 5. Fishes – Homing. I. Scholz, A.T. (Allan T.), 1948–. II. Title. III. Series. QL638.S2H26. 1983. 597'.55. 83-12467.

Softcover reprint of the hardcover 1st edition 1983

2131/3130-543210

Dedicated to the enthusiastic, hardworking and innovative graduate students, staff and faculty at the University of Wisconsin, Center for Limnology, who collaborated in these studies. A.D.H.

and

Dedicated to my parents. A.T.S.

Contents

Foreword

Chance Favors Only the Prepared Mind

How does a scientist go about the task of pushing back the curtains of the unknown? Certainly the romance of tackling the mysteries of nature provides the motivation, for who would not be inspired by the remarkable life history of this romantic beast, the salmon.

After living in the Pacific Ocean for several years, salmon swim thousands of kilometers back to the stream of their birth to spawn. I have always been fascinated by the homing migration of salmon. No one who has seen a 20-kilogram salmon fling itself into the air repeatedly until it is exhausted in a vain effort to surmount a waterfall can fail to marvel at the strength of the instinct that draws the salmon upriver to the stream where it was born. But how does it find its way back?

I was puzzling over this problem during a family vacation in 1946. Inspired by the work of the great German Nobel Laureates, Karl von Frisch and Konrad Lorenz, I had been conducting research with my graduate student Theodore Walker, since 1945, on the ability of fishes to discriminate odors emanating from aquatic plants.

Von Frisch had studied schooling minnows and discovered that, if broken, their skin emitted a conspecific chemical substance, termed Schreckstoff, which caused other members of its school to disperse and hide. The biological significance of this scent and subsequent modification in behavior was clear: If one member of the school had been injured owing to attack by a predator, the chemical acted as a sign stimulus to signal a warning and to elicit a stereotyped pattern of behavior by other members of the school that moved them out of danger. In fact, the incident that first drew the attention of von Frisch to this problem was his observation of a Kingfisher swooping down and attacking a school of minnows. After successfully capturing one, as the bird was flying away it dropped the minnow and von Frisch noticed that the other minnows, which were still schooling, suddenly dispersed.

Konrad Lorenz had been investigating imprinting in geese. Imprinting is a process of rapid and irreversible learning during a critical period of development that generally elicits a stereotyped pattern of behavior. In geese, the critical period occurs shortly after birth, when the gosling forms a permanent attachment to the first moving object it sees – normally its mother. Lorenz proved his thesis by assuming the role of a surrogate mother. His adopted gosling, hatched artificially, forsook its avian ancestry and formed a bond with Lorenz as witnessed by the fact that the bird followed him around like a shadow everywhere he went. The biological significance of this stereotyped behavior is that it enables the juvenile to remain close to and, hence, protected by, its mother.

During my vacation an incident occurred that caused me to incorporate the imprinting concept with the problem of homing in salmon. As Louis Pasteur is reported to have said, "Chance favors only the prepared mind."

We had driven across the sage country and high desert from Madison, Wisconsin, where I had recently joined the Faculty of the Zoology Department, to my parental home in Provo, Utah. Philosophically, this is about as far away from salmon country as possible. As I hiked along a mountain trail in the Wasatch Range of the Rocky Mountains where I grew up, my reflections about the migratory behavior of salmon were soon interrupted by wonderful scents that I had not smelled since I was a boy. Climbing up toward the Alpine zone on the eastern slope of Mt. Tinpanogos, I had approached a waterfall which was completely obstructed from view by a cliff; yet, when a cool breeze bearing the fragrance of mosses and columbine swept around the rocky abutment, the details of this waterfall and its setting on the face of the mountain suddenly leapt into my mind's eye. In fact, so impressive was this odor that it evoked a flood of memories of boyhood chums and deeds long since vanished from conscious memory.

The association was so strong that I immediately applied it to the problem of salmon homing. The connection caused me to formulate the hypothesis that each stream contains a particular bouquet of fragrances to which salmon become imprinted before emigrating to the ocean, and which they subsequently use as a cue for identifying their natal tributary upon their return from the sea. I envisioned that the soil and vegetation of each drainage basin would impart a distinctive odor to the water, thereby providing the salmon with a unique cue for homing. Later, I formalized this hypothesis in collaboration with my student, Warren Wisby, in 1951.

This monograph deals with olfactory imprinting and homing, and with endocrine control of the olfactory imprinting process in salmon. With my graduate students and associates I have spent the better part of 35 years trying to resolve the olfactory hypothesis for salmon homing. The list of students and collaboraters is long and in chronological order includes Warren J. Wisby, Ross M. Horrall, Andrew E. Dizon, Aivars B. Stasko, Dale M. Madison, Jon C. Cooper, Peter Hirsch, Peter B. Johnsen, and Allan T. Scholz. Dr. Scholz has been close collaborator since 1966 as an undergraduate field and laboratory assistant, graduate student and postdoctoral fellow. The present account, a revised version of his doctoral dissertation, focuses upon our recent investigations of olfactory imprinting and homing in salmon. It principally covers studies conducted since my book on salmon homing, *Underwater Guideposts: Homing of Salmon*, which appeared in 1966. However, since each of my students built on the foundation of information laid down by his predecessors, this monograph also provides a chronological account or our investigations. Our aim in this monograph is to trace the evolution of our ideas and the stages by which our experimental approach for studying olfactory imprinting and homing in salmon has developed.

Madison, WI, Spring 1983 ARTHUR D. HASLER

Historical Perspective

In studying the sensory basis for salmon homing, many investigators focus on the return of adults. However, what happens during the early stages of the life history of a salmon is of critical importance to understanding the homing process. Of particular significance is smolt transformation – the developmental stage when juvenile salmon undergo a series of morphological, physiological, and behavioral transitions at the time of their seaward migration. Juvenile salmon that were transplanted from their natal tributary to a different river before smolt transformation returned as adults to the second river. Thus, these fish adopted the river of release as the home stream. This suggests that during smoltification, salmon become indelibly "imprinted" to some property of their natal tributary that later serves to identify it when they return as adults to spawn.

The olfactory hypotheses for salmon homing as proposed by Hasler and Wisby in 1951 states that: (1) during the smolt stage juvenile salmon become imprinted to the distinctive order of their natal tributary, and (2) adult salmon use this information, stored in long-term olfactory memory, as a cue to relocate the stream during the spawning migration.

Initially, Hasler and Wisby were able to demonstrate in the laboratory that rivers have unique odors that salmon can differentiate by the sense of smell. Moreover, in field studies conducted in 1954 they found that returning salmon, when displaced below the confluence of two streams, could not return to their home river if their nostrils were plugged. As their work was with adult fish, Hasler and Wisby recognized that their studies did not establish that salmon have a long-term olfactory memory of the home-stream odor, and realized that the only way to tests this particular aspect of the hypothesis would be to work with both smolts and adults. Hence, they proposed to test for long-term olfactory memory (i.e., imprinting) by substituting a synthetic chemical for the natural scent of a home tributary, imprint juveniles to it during the smolt stage, and add the chemical to a river in order to decoy spawning adults to the site where the chemical was introduced. The chemical would have to be stable in water, detectable in small quantities so as to reduce any potential damage to the stream system into which it was introduced, neither naturally repellent nor attractive to salmon, and not normally found in the natural environment. The problem of selecting an appropriate compound was given to Warren Wisby for his doctoral research. His tests demonstrated that an organic compound, morpholine, would be suitable.

Because the stocks of salmon needed for the study were on the Pacific coast, while their working base lay 3000 km away in Wisconsin, Hasler and Wisby had published the details of their planned experiment in 1951, hoping that someone else would perform it. Although several scientists from other universities and government agencies expressed interest, the experiment was never performed. Then, in the late 1960's the Departments of Natural Resources in Michigan and Wisconsin introduced coho salmon into the Great Lakes to reduce the alewife population and to revitalize the Lake Michigan fishery. Suddenly, Hasler had, in his own back yard, the fish that he needed.

Through "the slings and arrows of outrageous fortune", not to mention a fortuitous coupling on the part of my parents during the "baby boom" that followed World War II, I arrived on the scene in Madison, at the same time as the salmon. In other words, in time to collaborate in performing the experiment proposed by Hasler and Wisby.

In a paper published in *Science* in 1976 we documented homing by coho salmon in Lake Michigan to two different synthetic chemicals. Our basic procedure was to expose (imprint) salmon smolts to synthetic chemicals, either morpholine or phenethyl alcohol, in the place of natural home-stream odors in order to determine if, as adults, they could be attracted to a stream scented with that chemical. The fish were exposed to the chemicals in a fish hatchery during the smolt stage and then stocked directly into Lake Michigan. During the spawning migration 18 months later morpholine and phenethyl alcohol were introduced into separate streams in order to simulate home streams for the experimental fish.

Results demonstrated that 95% of the morpholine fish were captured in the morpholine-scented stream and 92% of the phenethyl alcohol fish were recovered in the phenethyl alcohol-scented stream. We concluded that, since the experimental procedure ensured that they could not learn alternate cues for identifying the test streams, the fish were attracted to the streams by the synthetic chemicals. The fish were able to learn, or imprint to, the chemicals during a brief period of time at the smolt stage, and retain these cues for 18 months without being again exposed to the chemicals. As this study was conducted in the field, it provides direct evidence that coho salmon use this mechanism for homing.

Recently, in collaboration with Robert W. Goy, we have begun to investigated how smolt transformation and the olfactory imprinting process are regulated (summarized in Scholz 1980). In these studies plasma levels of thyroid hormones and cortisol were measured by radio-immuno-assay at approximately 4-week intervals in coho salmon from the pre-smolt stage in January through smolt transformation in May. At each interval morphological, physiological and behavioral transitions were quantified. Plasma levels of thyroid hormone and cortisol were found to increase at the end of April just before development of the transitions that mark the smolt stage, i.e., increased silvery coloration, salinity tolerance, osmoregulatory capability, Na^+/K^+ ATPase activity in gills, and downstream migratory activity.

In a second experiment, pre-smolt coho salmon were injected with thyrotropin (TSH), adrenocorticotropin (ACTH), TSH+ACTH, saline solution (controls), or left uninjected (controls) for 3 weeks in January. Plasma levels of thyroid hormone and cortisol were monitored by radioimmunoassay. Before and after hormone treatment the degree of smoltification was assessed by measuring morphological, behavioral, and physiological transitions. The degree to which transitions in experimental fish actually resembled process occurring in natural smolt transformation was determined by quantitative comparisons between experimental fish and natural smolts. Pre-smolt fish injected with TSH, ACTH, or both, in concentrations sufficient to elevate plasma thyroid hormone and cortisol values to levels observed in natural smolts, exhibited various characteristics associated with smolt transformation, whereas control fish did not. Treatment with ACTH increased salinity tolerance and osmoregulatory capability. Treatment with TSH increased silvering and downstream migratory behavior. The physiological and behavioral transitions in hormone-injected pre-smolts resembled those seen in salmon undergoing natural smolt transformation. Since the physiological and behavioral transitions were not observed in saline and uninjected pre-smolts, we infer that the hormones, and not some other facet of the experimental procedure, accelerated smolt transformation. Thyroid hormones and cortisol seem to be causal and not just correlative factors in initiating the smoltification process in the sense that: (1) their concentrations increase just before the behavioral and physiological transitions occur, and (2) injection of TSH and/or ACTH into pre-smolts induces these transitions.

There is also evidence that olfactory imprinting occurred owing to the TSH injections. Pre-smolts receiving TSH and exposed simultaneously to synthetic chemicals later demonstrated the ability to track their respective odor upstream, whereas subjects receiving no injection, saline or ACTH, did not. In this study, fish were retained for 10 months after the period of hormone treatment and odor exposure. Odor-discrimination tests were conducted in a natural river system below the confluence of two tributaries. Morpholine and phenethyl alcohol were introduced into either branch, and fish were released 150 m below the junction. A response required upstream migration and selection of a tributary scented with the correct treatment odor. Hence, this experiment duplicates the kind of choice that must be made by naturally migrating adult fish. As an additional control measure natural smolts were exposed to morpholine or phenethyl alcohol as in our previous experiments and tested along with the experimental fish. This permitted evaluation of responses of experimental fish in comparison with "naturally imprinted" fish. The behavior of TSH-injected fish was similar to that displayed by these "naturally imprinted" salmon.

This monograph is organized into two parts. Part I covers general information about the life history of salmon, the olfactory hypothesis, and describes our recent work on artificial imprinting of salmon with synthetic chemicals. Part II deals with hormonal regulation of smolt transformation and olfactory imprinting in salmonids and other factors controlling salmon migrations.

Because for the past 35 years we and our colleagues have had occasion to witness and study the migratory behavior of salmon in the Great Lakes, in the Columbia River and San Juan Islands (Washington State), in the Skeena and Fraser River systems (British Columbia), off the coast of Alaska and Japan, in France, in the North Sea and the Baltic Sea, this monograph is also a collection of general impressions that we have formed about the life history, behavior, physiology, and ecology of salmon.

Upon his retirement in March of 1978, A. D. Hasler's 52 doctoral students, a "full deck" so to speak, held a commemorative celebration in his honor at the University of Wisconsin. At the meeting I had the good fortune to meet Warren Wisby, the first of Hasler's students to work on problems of salmon homing. After I summarized our recent work, Wisby raised his hand to express delighted surprise at how far we had carried the work that he and Hasler had initiated over 30 years earlier. Then, he asked me how old I was as the time of their first experiments dealing with the detection of morpholine by salmon. This question puts the work described here in its proper historical framework: Hasler and Wisby began to investigate how olfactory imprinting might operate in salmon homing in 1947 – the year before I was born!

In many respects Hasler's work has come full circle. For example, 40 years ago Hasler and Roland K. Meyer were among the first biologists in North America to study fish endocrinology in their investigations on the effect of pituitary extracts for inducing spawning in trout. They then became interested in other problems and left the development and refinement of these techniques to other endocrinologists. These refined and developed techniques, handed down from the original work of Hasler and Meyer, were essential for my dissertation research and enabled me to perform several of my experiments effectively.

Laboratory experimenters often express dismay over the potential difficulties of doing rigorous, controlled experiments in the natural environment. We trust that the following pages will convince skeptics that by careful planning, coordination of laboratory and field experiments, adequate repetition, and development of logical hypotheses, rigorous field tests can be made.

The work described here also reflects the rewards of discussing ideas with associates in different disciplines. It was Marvin Johnson, a colleague in the Department of Biochemistry at the University of Wisconsin, who after hearing of our difficulty in finding a suitable imprinting chemical, first suggested that Hasler try morpholine. Robert W. Goy, a member of the Psychology Department and Director of the University of Wisconsin Regional Primate Center, collaborated with us on the endocrinology studies. Edward M. Donaldson, Director, Nutrition and Applied Endocrinology Program, Canadian Fisheries and Marine Service, Vancouver Laboratory, British Columbia, became interested in our project and supplied us with salmon gonadotrophic hormone before it was commercially available.

Field projects of the scope documented in this monograph require the support and cooperation of numerous people. We thank and appreciate the contributions of our field crew: Kathy Hughes, Cheryl Gosse Goodman, Mark Muzi, Timothy Whitney, Kieth Kiehnau, Roderick Smith, Terrance Chapp, Stephen Lewis, Donald Dodge, Joan Lindberg, Gunther Sheffler, Joan Sheffler, Fritz Wegner, and Catherine Balme. For field assistance we are also indebted to Russell I. Daly, Donald Czekleba, and Ronald Poff (Wisconsin Department of Natural Resources); William Boes (Army Corps of Engineers in Kewaunee, Wisconsin); Sylvester Drzeweicki (South Milwaukee Water Filtration Plant, South Milwaukee, Wisconsin); Edward Mueller (Biology teacher, South Milwaukee High School); and Francis "Toot" Weniger (Commercial Fisherman, Algoma, Wisconsin). The smoothness of our field operations can be directly attributed to their help. Counting them as personal friends was, for us, one of the most rewarding experiences of this study. Additionally many associates, in the United States, Canada, Japan and Europe, too numerous to acknowledge individually, have expanded our knowledge through discussion and correspondence. We can only encourage the unstinting sharing of ideas, since we believe that this is essential for continued progress in the sciences.

We thank Cathi Codi at the University of Wisconsin and Dalila Rivas, Carol Harmon, Eleanor Best, and Denice Shook at Eastern Washington University for typing drafts of the manuscript; and Dawn Holladay at Eastern Washington University for preparing the graphics. We also thank Professor Donald S. Farner at the University of Washington, Department of Zoology in Seattle, for his collaboration in editing our original manuscript. Finally, we thank Dr. Horace Simms and Heather McKean at Eastern Washington University, Department of Biology, for their editorial comments, many helpful suggestions, and careful proofreading of the final draft.

The 1960's and 1970's were exciting times at the University of Wisconsin Laboratory of Limnology. We were able to solve many of the riddles of the migratory behavior of salmon. Nevertheless, in any good research program more questions are generated than answered. Hence, before me on my desk at Eastern Washington University, Department of Biology, is a list of 124 questions about which I am still pondering on the migration of salmon. In collaboration with my colleague Ronald J. White, an endocrinologist at E.W.U., and our graduate students, I am presently trying to answer some of these questions about thyroid influences on imprinting behavior and brain maturation in salmon and trout. I expect to answer as many as three of four of them before I complete my career 40 years down the road and in the process generate at least 100 additional problems for future students to tackle.

Cheney, WA, Spring 1983 ALLAN T. SCHOLZ

Part I
Olfactory Imprinting and Homing in Salmon

Arthur D. Hasler and *Allan T. Scholz*

Chapter 1

Notes on the Life History of Coho Salmon

This chapter provides an overview on migration and homing of salmon, sketches the principal developmental stages in the life cycle of coho salmon (*Oncorhynchus kisutch*), and presents the evidence for imprinting.

1.1 The Magic Journey: Migration of Salmon

Salmon are euryhaline, anadromous fishes able to survive in, and make migrations between, fresh and salt water. Born in freshwater, they migrate to their feeding grounds in the ocean and back again for the purpose of spawning. The spawning migration of salmon is a magic journey: Their lives end in the same place where they began; their decaying bodies fertilize the drainage basin and, thus, indirectly generate the nutrients utilized by their offspring before the juveniles emigrate to the sea.

The onset of their migration is precise and regular, occurring at the same time every year. Once underway they adhere to a rigid schedule, arriving at certain points along their migration route on approximately the same date. In fact, their migration is so predictable that archeological evidence, obtained from pictures of salmon painted on the walls of caves in Spain, suggests that primitive cultures constructed their calendar year around the return of the salmon.

The spawning migration of salmon covers long distances, often 2,000–4,000 km. In many cases, the salmon spawn several hundred kilometers from their ocean feeding grounds in shallow mountain streams that form the headwaters of larger rivers. Their urge to swim upstream is impressive. During their upstream migration salmon traverse rapids and negotiate waterfalls to reach their spawning grounds. With their backs sticking out of water and tails vibrating frantically, they wriggle across gravel riffles. At the end of their arduous journey, the salmon spawn and die, leaving their progeny to migrate to the ocean, grow to maturity, and repeat the cycle.

1.2 Homing

Even more remarkable than their persistence is the salmons' homing "instinct", i.e., their ability to return almost unerringly to the same tributary in which they were born. Homing of salmon is well documented (reviewed by Hasler 1966, Harden-Jones 1968), being characteristic of five species of Pacific salmon – pink (*Oncorhynchus gorbuscha*), chum (*O. keta*), coho (*O. kisutch*), sockeye (*O. nerka*), and chinook (*O. tschawytscha*) – and also of Atlantic salmon (*Salmo salar*), rainbow trout (*S. gairdneri*), brown trout (*S. trutta*), and cutthroat trout (*S. clarki*).

The methods of studying homing behavior follow a general pattern. Young salmon are captured in their natal tributary before their seaward migration, marked with either a fin clip, external tag, or a magnetic wire implant and released back into the river where they were caught. The fish subsequently migrate downstream and, depending on the species, spend from 1 to 4 years in the ocean. During the spawning migration, the natal tributary is monitored for marked fish. Neighboring tributaries are also surveyed to determine if straying occurs.

According to Izaak Walton in *The Compleat Angler*, homing was first observed in Atlantic salmon (*Salmo salar*) in 1653 "by tying a ribbon or thread to the tails of young salmon which were swimming toward salt water and catching them again when they came back to the same place upon their return from the sea." Since then, scores of studies on the homing behavior of different species of salmonids have been conducted: In most cases, the return of fish to their natal tributary was precise (reviewed by Scheer 1939, Hasler 1966, Harden-Jones 1968). The results of these studies are remarkably consistent and indicate that, because of high mortality in the ocean, only about 0.5%–5% on the original downstream migrants survive to spawn and, of these, about 95% return to their natal stream, with the remainder straying into other streams.

Marked salmon caught at sea during the oceanic portion of their life cycle confirm that the fish migrate to distant places, and in a few instances biologists have actually obtained evidence of the total migration from river to sea and back again by capturing, in the ocean, a fish bearing a unique mark from a specific tributary, marking it a second time and releasing it back into the sea, and, finally, recovering the fish during the spawning migration in the tributary where it had been originally marked. The first example of this remarkable series of recaptures was reported by Huntsman (1942). Interestingly, at that time Hunstman was one of the leading opponents of the home-stream theory and did not believe that salmon range far away from the stream of their birth. It seems fitting that the first definitive evidence in support of the home-stream theory was provided by a biologist who was a skeptic. His observations were:

May 1938: Juvenile Atlantic salmon were marked during their descent of the Northeast Margaree River on Cape Breton Island, Nova Scotia.

June 1940:	One of these fish was caught by a commercial fisherman operating in the Atlantic Ocean off the coast of Newfoundland, about 1,100 km from the Margaree River. The fisherman, a Canadian government employee, noted and recorded the marked individual. Then, after marking the fish a second time, he released it back into the sea.
September 1940:	The fish was captured again in Northeast Margaree River where it had returned to spawn. The fish was caught by an angler who returned the tag to the appropriate authorities.

In a second example, a sockeye salmon, originally tagged during the smolt migration in the Thompson River, British Columbia, was later recaptured in the North Pacific Ocean, and ultimately recovered in the Thompson River during the spawning migration (Pritchard 1943). Delacey (1966) has also reported examples of coho salmon and steelhead trout that swam far away (753–2,700 km) from any possible influence of the parent river and returned.

1.3 Functions of Homing

An important consequence of homing is that it reduces reproductive wastage, i.e., spawning where conditions are unfavorable to the survival of eggs or juveniles, and ensures that spawning will be confined to waters of proven suitability for survival – proven in the sense that the spawners themselves were born and survived there.

Straying is probably as important as homing to the population, as it would maintain gene flow, thereby reducing problems of inbreeding. Straying also ensures that some members of the population will survive a catastrophe such as contamination or blockage of the home stream, and that newly available habitats will be colonized.

There is evidence that homing plays an important ecological role in stabilizing the freshwater ecosystems in which salmon spawn. There is constant loss of nutrients from salmon rivers to the ocean. Nutrients converted into salmon flesh during residence in the ocean are carried back upstream and recycled through decomposition of carcasses. The evidence for this conjecture comes from studies on the limnology of salmon rivers in British Columbia (reviewed by Foerster 1968), and especially by the elegant studies of Krokhin (1975) on the nutrient budgets of lakes in Kamchatka in which he estimated inputs and outflows of nutrients in nursery lakes of sockeye salmon. The latter estimated that decomposing salmon contribute 10%–

80% of the total annual input of nitrogen and phosphorus. The total annual budget of these nutrients depends on the number of spawning adults. By trapping adults in weirs, he found that the number (or biomass) of adult fish returning to spawn in the fall each year was different. Phosphate and nitrate levels of lake water measured the following spring correlated directly with the number of salmon returning: when large numbers returned more nutrients were present than when small numbers returned. Primary production and zooplankton populations were also larger in the spring after a large return. Since nutrient and production levels reflected the abundance of adult spawners, it is clear that decaying salmon bodies fertilize the watershed. The significance of these observations is that salmon play a central role in the ecosystem of recycling nutrients from the sea.

In these studies the number of young salmon that migrated downstream the following year also correlated with the number of returning adults: maximal numbers of emigrating juveniles occurred 1.5 years after maximal numbers of spawning adults. This pattern occurs because the increase in availability of food during the summer growing season immediately following a large autumn spawning run reduces competition among juveniles. Hence, more juveniles survive to emigrate to the ocean as 18-month-old smolts. Experimental fertilization of nursery lakes of sockeye salmon in British Columbia has increased the number of seaward migrants. The significance of these observations is that in the process of dying, adult salmon contribute to the nutrition and survival of their own offspring.

Migration and homing of salmon appear to constitute an energy-efficient, recycling system that operates at the ecological level of biological organization. Like the Krebs cycle, which is an energy-efficient recycling system that operates at the cellular level, the example of the salmon explains how complex structural organization and behavior evolve in apparent contradiction to the Second Law of Thermodynamics.

1.4 Life History of Coho Salmon

Coho salmon spawn in late October and November in streams that flow rapidly over gravel bottoms. After a ritualized mating dance with her breeding partner, the adult female turns on her side and scoops out a nest, or redd, by displacing gravel with her tail fin. After the eggs are laid and fertilized, she covers them with gravel. Exhausted by their effort, both parents die and their bodies drift downstream.

1.4.1 Alevin Stage

Incubation of the eggs and much of the early development occurs under the gravel. The eggs hatch about 1 month after deposition. The hatchlings, or alevins, remain buried for 4 months in the gravel, which protects them from predators. Alevins are nourished by a yolk sac, which contains a balanced diet of nutrients. A flow of clean, well-oxygenated water is critical to their survival. When the oxygen supply is reduced, e.g., if silt covers the redd, the alevins may suffocate.

1.4.2 Fry and Fingerling Stages

After their yolk sacs are absorbed, the alevins emerge from their redd. At this stage the young salmon, or fry, display territorial behavior and defend their feeding territory against encroachment by chasing and nipping at intruders. Stationing themselves underneath on overhanging bank, with their heads pointed upstream, they wait for their prey (mostly plankton or aquatic insects) to drift downstream with the water current.

The fry are themselves prey for larger fish and birds. Although their body coloration, dark green with distinct black vertical bars, or parr marks, blends in with stream bank vegetation and helps to protect them from predation, only about 100 of the 3,000 eggs laid by an adult female survive through the fry stage (Foerster 1968).

The fry remain in their natal stream until they are 18 months old during which time they grow in length from 3 to 10 cm. Salmon fry are also known as fingerlings when they reach finger-sized lengths, or as parr because of their distinguishing parr marks.

1.4.3 Transformation from Parr to Smolt

When they are 1.5 years old, in April or May, coho salmon undergo a metamorphosis referred to as smolt transformation, which involves a series of characteristic changes in morphology, physiology, and behavior (reviewed by Hoar 1976). During the smolt stage, their parr marks disappear, and they turn silver (adaptive coloration for living in the ocean); their osmoregulatory mechanisms begin adjustments that will enable survival in salt water; they cease territorial behavior, form schools numbering in the thousands of individuals, and embark, en masse, on their seaward journey. Just prior to the smolt stage, the endocrine system undergoes major transitions (reviewed by Hoar 1976). Smolt transformation is thought to be induced by thyroid hormones. (For details see section on the endocrine system in Chap. 3.) In the literature several terms have been used to describe this process, including smoltification, smolt transformation, parr–smolt transformation, and smolt stage. We prefer the term smolt transformation because it conveys the impression of a dynamic, metamorphic process as opposed to a static life-history stage. The dynamic nature of this process makes it exceedingly difficult to determine exactly when a salmon becomes a smolt and the opinions of any two experts may be radically different. Throughout the smolt transformation salmon are continuously changing from their freshwater fry stage into a subadult ocean stage.

The smolt stage is of critical importance to understanding the homing process because it is during this period that salmon become indelibly "imprinted" to some property of their natal tributary that serves later to identify it when they return as adults to spawn. Awareness of the home stream appears to be acquired rather than inherited insofar as young salmon transplanted from their natal tributary into a second stream a few days before smolt transformations occurs will return to that second stream where smolt transformation actually took place (see Sect. 1.5 for details).

1.4.4 Ocean Life: Migration to the Spawning Site

For 1.5 years coho salmon remain in the ocean, where they feed on squid, plankton, and small fish such as herring and alewives. While at sea they grow from 10 to 85 cm in length and gain from 29 g to 5 kg in weight. They spawn in autumn almost invariably at 3 years of age.

The spawning migration begins in midsummer as transitions in morphology, physiology, and behavior occur. The gonads undergo differentiation and a migratory disposition develops. Again, the endocrine system plays a key role in initiating these transitions (Poston 1978; see Chap. 5 for additional details). In early June the gonad mass is very small in both males and females (<0.1 kg for a 4.7 kg animal). By the end of August the gonad mass is nearly half the body weight of the animal (2.4 kg for a 4.7 kg animal). In migration, the fish stop feeding and mobilize fat reserves.

The homing migration consists of two main phases (Hasler 1966, Harden-Jones 1968): (1) an ocean phase, or open-water migration, when salmon migrate from the open water of oceans, or lakes, into shore areas near the home-river systems, and (2) a stream phase, or upstream migration, when they locate their main river and home tributary. Experimental evidence (reviewed by Hasler 1966, Harden-Jones 1968, Hara 1970) suggests that the sensory cues used in ocean migration and upstream migration are different (see Chap. 2 for details).

As the fish migrate upstream secondary sexual characteristics appear – their sides turn bright red, their upper jaws become hooked, and their gonads ripen. By the time of spawning the gastro-intestinal tract is completely resorbed. If a fish is cut open there is no trace of visceral organs except for the heart and kidney, and the entire body cavity is filled with ripe eggs or milt.

After spawning, the adults die, leaving their progeny to migrate downstream, grow to maturity in the ocean, and repeat the cycle.

1.5 Evidence for Imprinting: Transplantation Experiments

Studies concerning fish transplanted from their natal tributary to a different one before, during, or after the smolt stage, suggest that imprinting occurs during the smolt stage (reviewed by Ricker 1972).

1.5.1 Transplantation Before or During the Smolt Stage

When Rounsefell and Kelez (1938) transferred marked, presmolt coho salmon from their native river to another, the fish migrated to sea and returned as adults to the river into which they had been transferred. In addition, coho, chinook, sockeye and atlantic salmon or rainbow and brown trout, raised in a hatchery and transplanted before undergoing smolt transformation, returned to the river of release (Shapovalov and Taft 1954, Donaldson and Allen 1957, Lindsay et al. 1959, Carlin 1968, Shirahata and Tanaka 1969, Ellis 1970, Jensen and Duncan 1971, Fessler 1974, Aho 1975, Mahnken and Joyner 1975, Mighell 1975, Vreeland et al. 1975, Wahle 1975, Heard and Crone 1976, comprehensive review by Ricker 1972).

There is evidence that this process is rapid. Shirahata and Tanaka (1969) found that transplanted smolt-stage sockeye salmon returned to the stream of release. They believe that the time necessary for imprinting was less than 10 days, since the fish had left the river by that time. Carlin (1968) and Jensen and Duncan (1971) transplanted Atlantic salmon and coho salmon just as they began to smolt: The fish left the river within 2 days and returned to it during the spawning migration. Particularly illuminating was a study by Mighell (1975) who found that 4 h was a sufficient period for holding coho smolts in a new stream to ensure homing to it.

In all of these experiments the rate of return was observed to be about 0.5%–2% (ocean runs) or 2%–5% (freshwater runs), with about 95% of the recovered fish captured in the stream of release, which compares favorably with homing in natural populations.

1.5.2 Transplantation After Smolt Transformation

In contrast, when Peck (1970) transplanted hatchery-raised coho salmon to a Lake Superior tributary several weeks after smolt transformation, the return to the stream of release was poor (0.25%) with many recoveries in other streams. Peck concluded that the fish may have spent the sensitive period in the hatchery, where the water supply was not connected with Lake Superior, and suggested that imprinting terminates soon after smolt transformation begins, thereby preventing fish from becoming imprinted to other tributaries during the course of their downstream migration.

Additional support for this view comes from a study by Stuart (1959) with brown trout at Dunalastair Reservoir in Scotland. In this land-locked population, the fish grow to maturity in the reservoir and spawn in the tributaries. Stuart transplanted one group of brown trout from their natal tributary to a different one before smolt transformation had begun, while a second group was retained through the smolt stage before being transferred. Fish transplanted before undergoing smolt transformation returned to the river of release to spawn, whereas those transplanted after smolt transformation returned to their natal tributary.

In efforts to improve stocking techniques to maximize the ultimate return of stocked fish, the time of release has been found to be prominent in achievement of success (Reimers 1979). There is evidence that releasing hatchery-raised fish too early before or too late after smoltification reduces the return to the site of release.

Ellis (1957) found that Chinook Salmon transplanted after smoltification in the hatchery returned to the river where the hatchery was located instead of to the stocking site. Additionally, King and Swanson (1973) and Winter (1976) found that the time at which transplanted fish are stocked might influence homing to the stream of release; straying of rainbow trout in Lake Superior was greater, for example, if the fish were released late (July – as post-smolts) than if released early (April – just before smolt transformation). Reingold (1975a, b) made similar observations on rainbow trout in the Snake River, Idaho, and obtained comparable results.

1.5.3 Transportation of Smolts

The interpretation of Peck has been supported by investigations on coho, chinook, and rainbow smolts in the Clearwater–Snake–Columbia River system in Idaho, Washington, and Oregon (Ebel 1970, Ebel et al. 1973, Park 1975, Slatick et al. 1975, Weber 1975, reviewed by Collins 1976). Coho, chinook, and rainbow smolts were collected and marked after they had migrated downstream 120–140 km from their natal tributaries in the Clearwater and Snake Rivers, and then transported 570 km downstream to the mouth of the Columbia River (Fig. 1.1). During the spawning migration fish were recovered mainly at their natal streams; very few were recovered from the transplant site or other locations in the Columbia River system. Fish held at the Dworshak Hatchery on the Clearwater River until they had completed the initial stages of smolt transformation and then transported 810 km for release in the lower Columbia River were subsequently recovered at the hatchery 18 months later (Weber 1975). None were recovered at downstream monitoring stations. Similar results were obtained in a comparable experiment with Atlantic salmon (Mills and Shackley 1971).

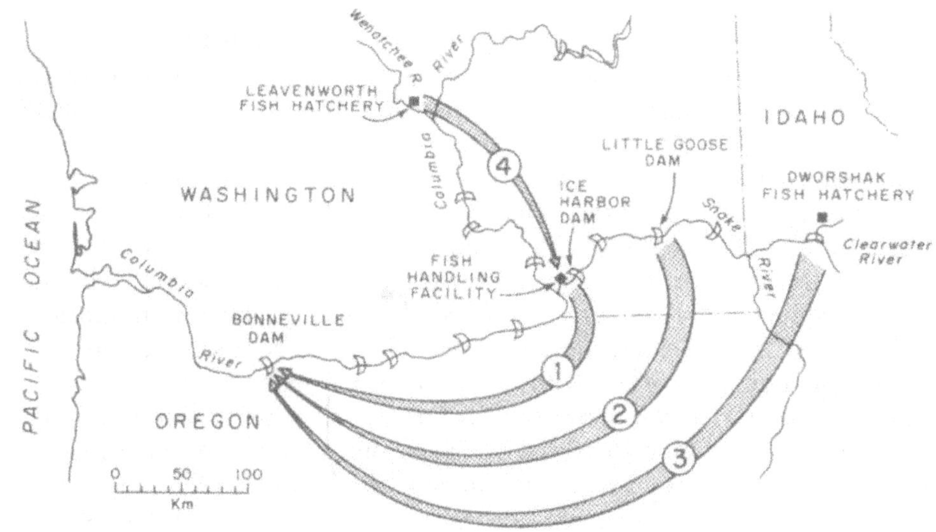

Fig. 1.1. Routes used for smolt transportation studies on the Columbia River

1.5.4 Coincidence of Imprinting and Smolt Transformation

Two important conclusions can be drawn from these transplantation studies: (1) the memory of the home stream is not inherited, and (2) homing is connected with a period of rapid and irreversible learning, i.e., imprinting, of the cues that identify the home stream at the time the young salmon begin their downstream migration. Regarding the first point, however, Bams (1976) has argued that there is also a genetic component. To test this hypothesis, he transplanted pink-salmon eggs from their original tributary (donor stream) to a second one (recipient stream). One group of eggs was cross-fertilized by males from the recipient stream; one group was simply transplanted. Both groups were raised in the recipient stream and then marked and released. About equal numbers of both groups left the recipient stream, but only about half as many from the pure donor stock as from the hybrid stock returned to it. Bams concluded that "imprinting alone brought back some of the pure donor stock," and "addition of the local male genetic complement improved the return to the river of release."

1.5.5 Stocking of Salmon in the Great Lakes

A by-product of the transplantation experiments has been that they have provided the basis for artificial propagation and introduction of salmon in areas where they are not native, for example in Lake Michigan, where most of our recent investigations were conducted. Streams that flow into Lake Michigan are not suitable for natural reproduction because of a lack of gravel substrates for building redds, heavy silt deposition, and poor water quality. Consequently, salmon runs in these streams are maintained predominantly by annual stocking of hatchery-raised fish. At 18 months of age coho salmon are transferred to "smolting ponds" in a tributary where they smolt and from which they migrate downstream to the Lake. Of the fish originally stocked 2%–8% are recovered during the spawning migration and, of these, 95% are recovered in the stream of release. We released 10,000 coho salmon marked with distinctive fin clips into three different Wisconsin tributaries of Lake Michigan, and during the spawning migration 18 months later monitored the three streams, along with 17 additional rivers flowing into Lake Michigan, for marked fish (i.e., all of the rivers flowing into Lake Michigan along the Wisconsin shoreline). Of the fish released in the first tributary 837 were recovered there and 18 at other locations. Of the fish released in the second tributary, 749 were recovered in that river and 26 from other locations. Of the fish released in the third tributary, 518 were recovered in that river and 14 from other locations (Cheryl Gosse Goodman and A. Scholz, personal observations).

1.5.6 Summary of Transplantation Experiments

The experiments with fish transported downstream after smolt transformation clearly suggest that they became imprinted to some factor in the water – that is, fish found their way through the home-river system and reached their natal stream without having used that route for the downstream migration. The main drawback with most transplantation experiments, however, is that they do not identify the cue to which the fish become imprinted. Our own work has mainly been concerned with trying to determine whether odor acts as the cue.

Chapter 2

Imprinting to Olfactory Cues: The Basis for Home-Stream Selection by Salmon

The olfactory hypothesis for salmon homing, first presented by Hasler and Wisby in 1951, has three basic tenets: (1) because of local differences in soil and vegetation of the drainage basin, each stream has a unique chemical composition and, thus, a distinctive odor; (2) before juvenile salmon migrate to the sea they become imprinted to the distinctive odor of their home stream; and (3) adult salmon use this information as a cue for homing when they migrate through the home-stream network to the home tributary. This chapter reviews the evidence for olfactory imprinting in salmon. Two central themes are intertwined throughout this section: (1) the results of experimental replications performed by different investigators are remarkably consistent, and (2) many experiments were conducted in the field instead of the laboratory and thereby provide direct evidence that salmon use olfactory cues for homing.

2.1 Early Investigations of the Olfactory Hypothesis

Actually, speculation about the use of odors as a cue for homing by salmon dates back to the 19th century (Trevanius 1822) and has been resurrected several times (Buckland 1880, Craigie 1926); but the idea was not refined until Hasler and Wisby formulated their hypothesis (Hasler and Wisby 1951). Moreover, Hasler and Wisby indentified several conditions about the hypothesis that could be tested, and proceeded to find ways to test them. They speculated that the use of olfaction for homing requires that: (1) each stream must have a characteristic and persistent odor perceptible by the fish; (2) the odor must have significance only to those salmon that originated in that stream: any odor that is a generalized attractant or repellent, which would induce all salmon to enter or reject a stream, would not be suitable, i.e., it would not be unique; (3) fish must be able to discriminate among the odors of different streams; (4) fish must be able to retain an "odor memory" of the home stream during the period intervening between the downstream migration and homing migration.

2.1.1 Laboratory Conditioning Experiments

The first objective was to determine whether or not fish could discriminate between two streams by smell (Hasler and Wisby 1951, Wisby 1952). Using reward (food) and punishment (electric shock) for conditioning, they trained groups of bluntnose minnows and coho salmon to discriminate between waters collected from two streams. When their nasal sacs were cauterized, the trained fish were not able to discriminate between the waters, indicating that it is the characteristic odor of each stream that is discernible by fish. In addition, they found that trained fish were not able to identify a water sample if the organic fraction was removed. Also, fish trained to discriminate water obtained from a stream during one season were able to discriminate water from the same stream collected in a different season, which suggests that the factor detected by the fish was long-lasting and present in the stream throughout the year. This, they felt, was an important point, because salmon may reside in the ocean for several years before returning to their natal stream, so the imprinting factor must be long-lasting. This experiment was repeated with similar results by other investigators (McBride et al. 1964, Tarrant 1966, Walker 1967, Jahn 1976).

2.1.2 Sensory-Impairment Experiments

The second important contribution by Hasler and Wisby was their performance of the first systematic field test of the olfactory hypothesis (Wisby and Hasler 1954). They recognized that while their laboratory studies provided information about the perceptual capabilities of salmon, the results obtained were not biologically meaningful in terms of ascertaining whether or not the fish actually use odor cues for homing. As a result, they decided to devise a field experiment. Their research site was a small Y-shaped tributary located 25 km from Seattle, Washington: Issaquah Creek and its East Fork. Each branch had its own native stock of coho salmon (Fig. 2.1).

Hasler and Wisby captured fish in both tributaries during the spawning season. The nasal chamber of half the fish from each tributary was plugged with vaseline-coated cotton, while the remaining fish were left unplugged to control for olfactory impariment. Each fish was tagged so that it could be identified in terms of the treatment it had received and the branch where it was caught. All of the fish were then released 1.6 km below the fork of the stream and allowed to repeat their upstream migration. Of those recaptured, all but 8 of 73 control fish returned to the branch where they were originally captured, but 28 out of 70 fish that had been deprived of their sense of smell entered the wrong branch of the stream, demonstrating that

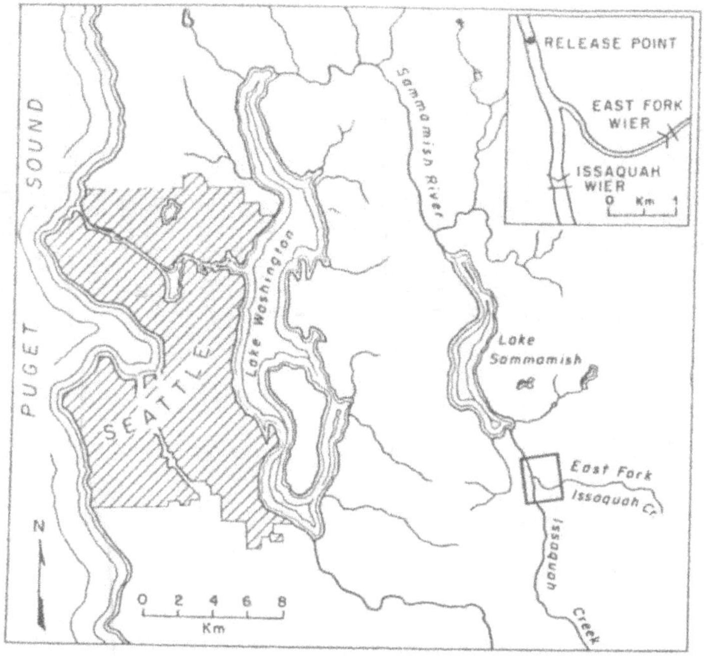

Fig. 2.1. Study area for olfactory-impairment experiments in Lake Washington watershed, adapted from Wisby and Hasler (1954). *Inset* shows detail of study site at Issaquah Creek and its East Fork

fish deprived of their olfactory sense exhibited a random pattern of movement.

This type of sensory-impairment experiment has been repeated 20 times by other investigators (Table 2.1). Species tested include coho, chum, chinook, and Atlantic salmon, and cutthroat, rainbow, and brown trout. The results are remarkably consistent and agree closely with the findings of Hasler and Wisby: In 16 experiments the olfactory sense appeared to be necessary for correct homing. In addition, two studies demonstrated that blinded fish homed nearly as well as control fish, and, thus, that vision was not essential for relocating the original stream, at least during the upstream migration (Hiyama et al. 1966, Groves et al. 1968).

Sensory-impairment experiments were criticized by Brett and Groot (1963), Harden-Jones (1968), Peters (1971), and Ramsey (1961), who argued that nose-plugged fish may home with less precision because of generalized traumatic or inhibitory effects rather than because of loss of the olfactory sense. Consequently, the results of sensory-impairment experiments are difficult to interpret.

Another problem with ablation experiments is that the fish had been exposed to their home water shortly before being tested, so the homing be-

Table 2.1. A summary of sensory-impairment experiments conducted with salmonids and other species of migratory, anadromous fishes

Species	Nasal sac occluded	Olfaction necessary for homing	Vision impaired	Vision necessary for homing	Investigators
Sockeye salmon	Yes	Inconclusive	No		Cragie (1929)
Coho salmon	Yes	Yes	No		Wisby and Hasler (1954)
Brown trout	Yes	Possible [a]	Yes	No	Stuart (1957)
Brown trout	Yes	Yes	No		Shearer (1959)
Chinook salmon	Yes	Yes	Yes	No	Collins et al. (1962)
Sockeye salmon	Yes	Inconclusive	No		Hartman and Raleigh (1964)
Sockeye salmon	Yes	Yes	Yes	Inconclusive	Lorz and Northcote (1965)
Chum salmon	Yes	Yes	Yes	No	Hiyama et al. (1966)
Cutthroat trout	Yes	Yes	No		Jahn (1966)
Chum salmon	Yes	Yes	Yes	No	Sato et al. (1966)
Cutthroat trout	Yes	Inconclusive [b]	Yes	Inconclusive [b]	McCleave (1956)
Chinook salmon	Yes	Yes	Yes	No	Groves et al. (1968)
Chinook salmon	Yes	Yes	No		Delacy et al. (1969)
Cutthroat trout	Yes	Yes	Yes	No	Jahn (1969)
Sockeye salmon	Yes	Yes	No		Shirihata and Tanaka (1969)
Cutthroat trout	Yes	Inconclusive [a,b]	Yes	Inconclusive [b]	McCleave and Horrall (1970)
Cutthroat trout	Yes	Yes [c]	No		LaBar (1971)
Cutthroat trout	Yes	Yes	Yes	Yes	McCleave and LaBar (1971)
Shad	Yes	Yes	Yes	Yes	Dodson (1973)
Shad	Yes	Yes	No		Legget et al. (1973)
Brown Trout	Yes	No	No		Winter (1970)

[a] Some problem with nose plugs. Most experimental fish recaptured in the homestream had missing plugs
[b] Both groups of treated fish homed as the controls but more slowly than control fish
[c] Some of the experimental fish homed but took more time than the controls

havior exhibited during the experiment may be owing to "short-term conditioning" rather than a "long-term memory" of the stream (Brett and Groot 1963).

2.1.3 Experiments on Olfactory Recognition of the Home Stream: Effects of Attractants and Repellents

Additional evidence from laboratory studies with adult fish captured in their home stream indicates that salmon can distinguish their home water from other waters (Idler et al. 1961, Fagerlund et al. 1963). Locomotor activity of salmon held in tanks of neutral water increased when water from the home stream was added, whereas water from other salmon-spawning streams did not produce this response. In a reciprocal experiment, salmon from a different tributary became more active when water from their own stream but not other streams was added.

Other studies have been carried out on the effects of some chemicals in blocking the upstream migration of salmon (i.e., repellent effects). Brett and MacKinnon (1953) measured the upstream progress of migrating salmon in relation to a variety of odorous substances poured in upstream. Water that had been in contact with mammalian skin (e.g., bearpaw, human hand rinse) acted as a strong repellent. Biochemical studies demonstrated that the repellent effect is owing to 1-serine in mammalian skin, i.e., purified 1-serine elicited a typical alarm reaction (Idler et al. 1961).

2.1.4 Electrophysiological Studies

Electrophysiological studies also suggest that salmon can distinguish their home water from other water (Hara et al. 1965, Ueda et al. 1967). Adult salmon were captured after arriving in their natal tributary and brought into a laboratory. Electroencephalograms from an electrode inserted into the olfactory bulb were recorded on a polygraph and the amplitude of response of each fish to water from different salmon spawning tributaries was measured. When the nasal sacs were flushed with water from the stream in which the fish was captured, a characteristically high-intensity EEG response was recorded. This response seemed to be specific because it was not evoked by water from other tributaries. Fish from several tributaries were tested and each responded most strongly to water from its own tributary.

Subsequent EEG studies (Oshima et al. 1969a–c, 1973, reviewed by Hara 1970, Ueda et al. 1971, 1973, Cooper and Hasler 1973, Dizon et al. 1973a, b, Bodznick 1975, Kaji et al. 1975, Satou and Ueda 1975, Hahn 1976) have generally confirmed the basic conclusions of earlier experiments. However, in some of this more recent work responses to some non-home water samples could not be distinguished from those to home water and occasionally elicited even stronger responses than did home water.

2.1.5 Criticisms of Early Investigations

Although most studies described so far have, in general, suggested that the olfactory response to water from the home stream is specific, Brett and Groot (1963), as mentioned above, have pointed out that since the fish used for olfactory impairment, home-stream discrimination, and EEG experiments had been exposed to home water shortly before being tested, the characteristic response displayed by the fish may have resulted from "short-term conditioning" rather than "long-term memory." Support for this argument comes from some EEG experiments. Oshima et al. (1969 b) demonstrated that salmon which had previously shown no olfactory-bulb response to water from the hatchery at the University of Washington College of Fisheries did respond after they had been held in this water for 67 h; in fact, they responded more strongly to this water than to their home water. Thus, since the fish used in most experiments described above were captured in the home stream, it is not clear if a response to home water in related to long-term olfactory memory of home-stream odors or to recent exposure to the water prior to testing.

Another criticism (see Sect. 2.1.2) is that in sensory-impairment experiments, nose-plugged fish may home with less precision because of generalized traumatic or inhibitory effects rather than because of loss of the olfactory sense. Also, in all of the experiments cited above only adult fish were used. Their imprinting history was not definitely known and it was assumed that the stream in which the fish were collected was their home stream.

These criticisms condense to a central point: To determine if olfactory imprinting occurs, a definitive experiment must encompass two periods in the life of the salmon, the smolt stage, when imprinting takes place, and adult spawning migration, when the fish must use their "long-term olfactory memory" of the home tributary as a cue for homing.

In 1951, Hasler and Wisby were aware of these points. Nevertheless, conditioning, sensory-impairment, home-stream recognition, and EEG experiments did show that each stream has a characteristic and persistent odor and that salmon can distinguish between the odors of different streams – supporting the first two tenets of the olfactory hypothesis. The consistency of the findings encouraged Hasler and Wisby to continue to study this problem. However, they had to agree that these studies did not provide conclusive evidence for olfactory imprinting and long-term olfactory memory; hence they modified their plan of research to test this particular point of the hypothesis.

2.2 Experiments with Artificially Imprinted Salmon

Hasler and Wisby (1951) proposed to imprint smolting salmon artificially with synthetic chemicals as a substitute for natural stream odors, and later, to use the scent to attract the fish to a different tributary. This experiment has been the culmination of 30 years of research by Hasler and his coworkers (Scholz et al. 1975, 1976, reviewed by Hasler et al. 1978). The method eliminated recent experience as a factor and allowed us to manipulate olfactory cues without interference with the fish's olfactory sense, so that problems with sensory impairment were avoided.

The first step in conducting the experiment, the subject of Wisby's doctoral dissertation in 1952, was to find a suitable chemical for imprinting the fish. The chemical had to be an organic compound, since the previous work of Hasler and Wisby had demonstrated that the identifiable component of the stream was contained in the organic fraction. In addition, it had to be highly soluble in water, chemically stable in the natural environment, not normally found in the natural waters, and one that would neither naturally repel nor attract the fish. Finally, Wisby felt that it was important for the chemical to be detectable in small quantities to minimize any possible damage to a natural stream system. Wisby screened several substances and found an appropriate compound – morpholine (C_4H_9NO), a heterocyclic amine that could be detected by unconditioned coho salmon at a concentration of 5.7×10^{-10} M.

Dizon et al. (1973 a) conducted a preliminary electrophysiological experiment using this compound. Two groups of coho salmon were raised at a hatchery until the smolt stage, when one group was exposed to a concentration of 5.7×10^{-10} M morpholine for 1 month. The fish were retained at the hatchery for 10 more months before being brought into a laboratory for testing. Fourteen fish from each group were presented with 0.01% and 1% morpholine, and EEG's from the olfactory bulb were recorded. Fish that had been exposed to morpholine displayed significantly higher EEG activity than unexposed controls.

2.2.1 Experiments in Southern Lake Michigan – Imprinting Salmon with Morpholine

We then began conducting field experiments (Madison et al. 1973, Scholz et al. 1973, Cooper et al. 1976). In the spring of 1971, we transported 16,000

18-month-old coho fingerlings of the same genetic stock, hatched and raised under uniform conditions, from a State Fish Hatchery in central Wisconsin to holding tanks at a water-filtration plant in South Milwaukee. The fish were held there for 30 days during their smolting period in tanks supplied with water piped in from Lake Michigan; thus, during their entire early life history they were not exposed to water from any tributary of Lake Michigan.

A concentration of 5.7×10^{-10} M of morpholine was maintained throughout the exposure period in a tank containing half of the fish. The rest of the fish, held in a second tank which was not treated, served as a control group. The fish from each group were then marked with different fin clips and released into Lake Michigan 0.5 km south of the mouth of Oak Creek, the stream that was to be scented during the spawning migration (Fig. 2.2). We released both groups directly into the lake to reduce the possibility that they might learn additional cues about the test stream.

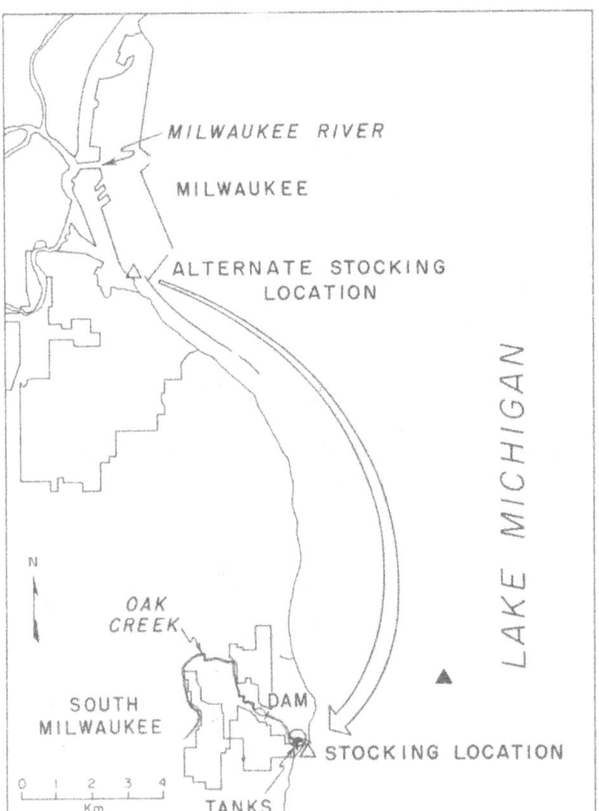

Fig. 2.2. Research area, South Milwaukee, Wisconsin, illustrating locations of: (1) the Oak Creek stocking site, and (2) the Milwaukee Harbor stocking site

Table 2.2. Census record of coho salmon caught at Oak Creek in fall 1972 (Scholz et al. 1973)

Treatment	Stocking location	Number released	Date released	Number recovered	Percent of fish stocked
Morpholine	0.5 km south of Oak Creek	8,000	May 1971	218	2.58
Control	0.5 km south of Oak Creek	8,000	May 1971	28	0.35

The fish remained in Lake Michigan for 1.5 years, then, during the spawning season in the fall of 1972 we created what we hoped would be an artificial home stream by continuous metering of morpholine into Oak Creek. We reasoned that if morpholine-exposed adult fish were attracted to Oak Creek it would probably be a result of the presence of morpholine in the stream, since, as juveniles, they had never had any experience with Oak Creek. The unexposed controls enabled us to determine if fish would return to the stream independently of the chemical cue. Adult salmon were collected in Oak Creek by electrofishing, gill netting, and creel census surveys. Because fish were unable to move past a dam situated 1.5 km from the mouth of Oak Creek, only a small portion of the stream had to be monitored. The creel surveys were made by walking back and forth along the bank from the dam to the river mouth and recording catches made by anglers. A complete circuit was made approximately every 1.5 h continuously from sunrise to sunset (the times when anglers were allowed to fish).

Electrofishing surveys were made to assess the number of fish not caught by anglers. We continued electrofishing until no fish were left in the stream. In view of the continuous monitoring and thorough census procedures, we feel that we saw most of the salmon present in Oak Creek.

Two hundred and eighteen of the fish that had been exposed to morpholine and only 28 from the control group were captured in Oak Creek in the fall of 1972 (Table 2.2). Since neither group had any previous experience with Oak Creek, these data suggest that the morpholine-exposed salmon became imprinted and were attracted to morpholine during the spawning migration. The recovery of morpholine-exposed fish represented 2.7% of those originally stocked, a result similar to those from homing studies conducted in the Great Lakes with natural or transplanted fish.

During the spring of 1972, we started experiments to replicate the series of 1971–72 (Scholz et al. 1975, Cooper et al. 1976). We worked with larger numbers of smolts; 18,200 were exposed to morpholine at a hatchery and 20,000 were left unexposed. This time we released the fish at two different points along the Lake Michigan shoreline, 0.5 km south and 13 km north

Table 2.3. Census record of coho salmon caught at Oak Creek in fall 1973 (Cooper et al. 1976)

Experi-ment	Treatment	Stocking location	Number released	Date released	Number recovered	Percent of fish stocked
1	Morpholine	0.5 km south of Oak Creek	5,000	May 1972	437	8.74
	Control	0.5 km south of Oak Creek	5,000	May 1972	49	0.95
2	Morpholine	0.5 km south of Oak Creek	5,000	May 1972	439	8.78
	Control	0.5 km south of Oak Creek	5,000	May 1972	55	1.10
3	Morpholine	13 km north of Oak Creek	8,000	May 1972	647	7.89
	Control	13 km north of Oak Creek	10,000	May 1972	65	0.65
TOTAL	Morpholine		18,000		1,515	8.42
	Control		20,00		169	0.85

Table 2.4. Census record of coho salmon caught at Oak Creek in fall of 1974 (Cooper et al. 1976). Morpholine was not present in Oak Creek

Treatment	Stocking location	Number released	Date released	Number recovered	Percent of fish stocked
Morpholine	0.5 km south of Oak Creek	5,000	May 1973	51	1.00
Control	0.5 km south of Oak Creek	5,000	May 1973	55	1.10

of Oak Creek (Fig. 2.2). In the fall of 1973, we caught 1,515 morpholine-exposed fish and 169 controls – a ratio of nearly 10 to 1, and over 8% return of the imprinted fish. Morpholine-exposed fish released as smolts 13 km north of Oak Creek homed to Oak Creek in a manner similar to those released near it (Table 2.3).

During a third (control) experiment, conducted in 1973, we exposed 5,000 smolts to morpholine and left an equal number unexposed. Again the fish were stocked 0.5 km south of Oak Creek, but in the fall of 1974 morpholine was not added to the stream. Exposed and control fish were captured in equally low numbers (51 vs. 55), at about the same rate as control fish from previous experiments (Table 2.4). This experiment illustrates the importance of morpholine to the return of imprinted salmon.

2.2.2 Experiments in Northern Lake Michigan – Imprinting Salmon to Morpholine or Phenethyl Alcohol

One major problem with these early artificial imprinting studies was that only the scented stream was monitored for marked fish, so it was possible that a significant number of morpholine-exposed fish were straying into other streams. In addition, we were not entirely satisfied that unexposed fish constituted an adequate control group. We thought that a better control would be to expose each of two groups of fish to a different chemical and then scent different streams with the chemicals, so that one chemical exposure would be the control for the other. Consequently, we decided to conduct a more refined experiment (Scholz et al. 1976).

Phenethyl alcohol ($C_8H_{10}O$) was used as an alternative chemical because Teichmann (1962) reported that rainbow trout could detect it at 4.1×10^{-8} M. One group of coho smolts held at a hatchery was exposed to morpholine (5.7×10^{-10} M), a second group to phenethyl alcohol (PEA) (4.1×10^{-8} M), and a third left unexposed. Each fish was given a fin clip which corresponded to the treatment odor it had received. All three groups were released in Lake Michigan midway between the two test streams, the Little Manitowoc River and Two Rivers, located 9.4 km apart (Fig. 2.3).

During the spawning migration, morpholine was metered into one test stream and PEA into the other. The streams were surveyed for marked fish by creel census, gill-net fishing, and electrofishing. In addition, 17 other locations (Fig. 2.3) were also monitored to determine whether a significant number of experimental fish were straying into non-scented streams. The effort spent in monitoring each location and the number of fish caught there is recorded in Table 2.5.

This experiment was conducted twice – the artificial imprinting was done in 1973 with 5,000 fish in each group and again in 1974 with 10,000 fish per group. The spawning migrations were in 1974 and 1975 respectively. The data from both experiments show that of the morpholine-exposed fish recovered, 95% were captured in the morpholine-scented stream; and, of the PEA-exposed fish recovered, 92.5% were captured in the PEA-scented stream (Table 2.5). By contrast, large numbers of control fish were captured at other locations.

The results demonstrate that olfactory imprinting occurs when the fish are 18 months old at the time they undergo the transition from parr to smolt; they retain this information for 1.5 years without again being exposed to the odor, and use it as an orientation mechanism to achieve successful homing.

As the work was conducted in the field, the results of our artificial imprinting experiments provide direct evidence for olfactory imprinting in coho salmon. We have also conducted three experiments with rainbow trout

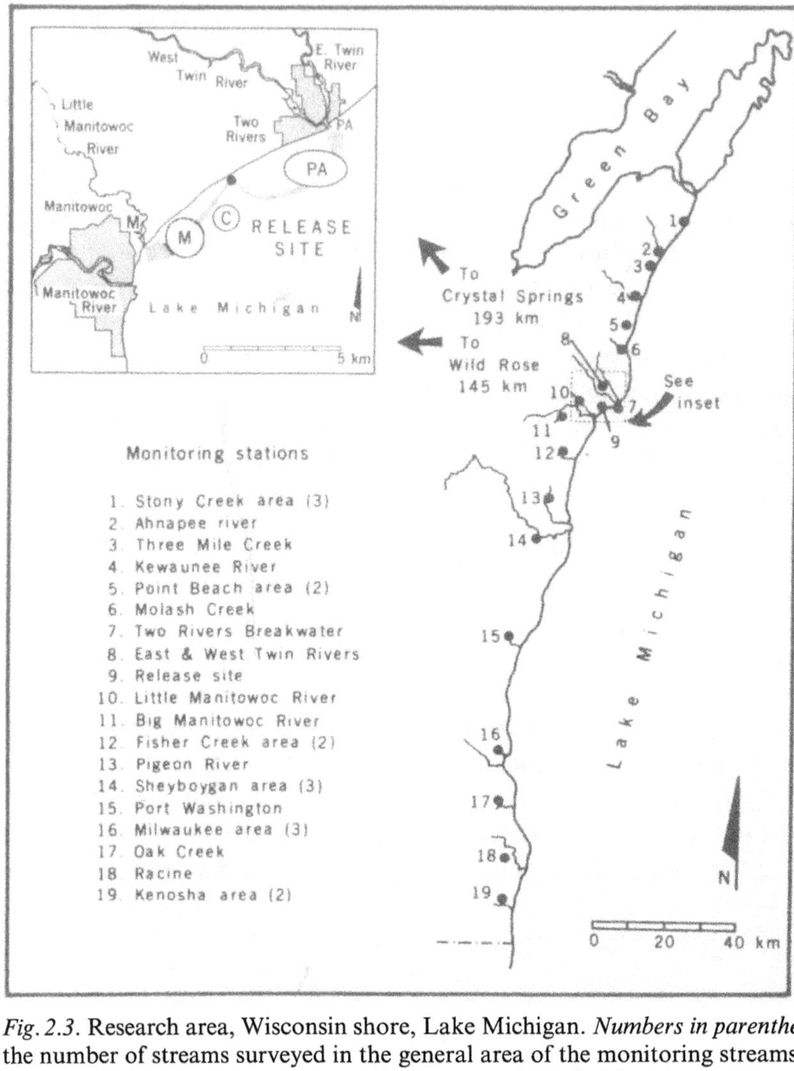

Fig. 2.3. Research area, Wisconsin shore, Lake Michigan. *Numbers in parentheses* represent the number of streams surveyed in the general area of the monitoring streams. *Inset* shows detail of the release site, the morpholine-scented Little Manitowoc River (*M*) and the phenethyl alcohol-scented breakwater area at Two Rivers (*PEA*)

(Scholz et al. 1975, 1978 a, Cooper and Scholz 1976) and one with migratory brown trout (Scholz et al. 1978 b) and in all cases significantly higher numbers of morpholine-exposed fish, as opposed to controls, returned to a stream scented with morpholine (ratios of 8–17 exposed : 1 control). In a preliminary experiment (Cooper et al. 1976), coho salmon which had been exposed to morpholine for 2 days at the onset of smolting returned to a simulated home stream in numbers about equal to fish exposed for 30 days.

Table 2.5. Total numbers of morpholine-exposed (M), phenethyl alcohol-exposed (PEA) and control (C) salmon captured at individual locations. Data from the Little Manitowoc River (morpholine-scented) and Two Rivers area (PEA-scented) are in boxes (Scholz et al. 1976)

Location			No. recovered in 1974			No. recovered in 1975		
			M	PEA	C	M	PEA	C
1.	Stony Creek (3)		1		4			12
2.	Ahnapee River			2	7	6	1	37
3.	Three Mile Creek		2	1	1			2
4.	Kewaunee River							
5.	Nuclear power plants (2)		1		4			2
6.	Molash Creek				2			
7.	Two Rivers breakwater	PEA	2	118	15	3	192	12
8.	East and West Twin rivers			15	7	3	8	21
9.	Stocking site		1		7			1
10.	Little Manitowoc River	MOR	207	6	24	452	14	52
11.	Big Manitowoc River		2	3	31		1	26
12.	Fisher Creek (2)				3			1
13.	Pigeon River							
14.	Sheboygan River (3)		1		3			3
15.	Port Washington							
16.	Milwaukee area (3)							
17.	Oak Creek			1	7			
18.	Racine				1			
19.	Kenosha (2)							

Thus, brief exposure to morpholine seems sufficient to imprint fish successfully.

In planning our experiments, we took into consideration several situations that could have biased the results by causing differences in the return of imprinted and control groups independent of chemical odor. These included:

1. Differences in genetic background and early life history between imprinted and control groups. In our experiments, both groups of fish had similar genetic backgrounds, being from eggs of fish taken at one location in Lake Michigan and raised under uniform conditions until separated.

2. Careless or incorrect marking of young fish. We assumed that any errors in fin clipping were equivalent for each group. The Great Lakes Fishery Commission in Ann Arbor, Michigan, assigns fin clips so that investigators working in Lake Michigan do not duplicate each other's fin clips.
3. Differential mortality because of different fin clips. To control for this possibility, paired groups received symmetrical clips. The incidence of mortality that occurred immediately after fin clipping was low (1 out of 1,000 fish) and similar for all groups. In addition, fish were clipped several months before the experiments were begun, to allow enough time for them to recover from the fin-clipping operation before being stocked in the lake. Therefore, we assumed that differential mortality did not bias the results. The possible mortal effects of fin clipping or other factors after the fish were released into the lake are not known.
4. Problems with identification because of regeneration of clipped fins. Since Rich and Holmes (1929) and Stuart (1958) reportet that double fin clips can reduce difficulties of identification, we used double clips in most experiments, further assuming that regeneration would be similar for each group.
5. Problems with misidentification of fin clips. To avoid this error, we worked in pairs with each person examining the fish independently.

2.2.3 Ultrasonic Tracking

In 1971–73 we conducted a different type of behavioral experiment to test the imprinting hypothesis. Essentially this involved releasing morpholine-exposed fish along the shoreline of Lake Michigan and tracking them into an area scented with morpholine (Scholz et al. 1973, 1975). Control experiments were conducted by tracking morpholine-exposed fish through the test area when morpholine was absent or when a different chemical was present, as well as tracking unexposed fish with morpholine present.

Adult salmon, captured at Oak Creek as part of the census experiments previously described at that site, were transferred (displaced) by boat to a point 3.2 km north of Oak Creek (Fig. 2.4). There, before releasing them, we inserted an ultrasonic transmitter down the esophagus into their stomach. A directional hydrophone connected to receiving equipment on a tracking boat was used to follow the signal from each fish as it swam along the shoreline.

We selected the release site assuming that fish released near the shore would follow the shoreline to Oak Creek and intercept an intervening point where a small stream flowed into the lake. Morpholine was introduced into the test area in a line perpendicular to shore, extending from the mouth of the stream to about 100 m offshore, creating a narrow band which acted

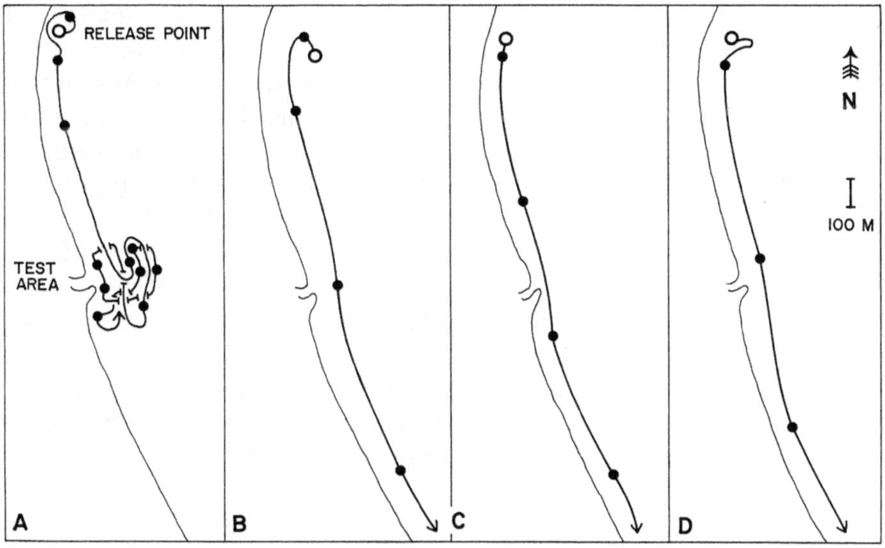

Fig. 2.4 A–D. Ultrasonic track movements of salmon through a test area that had been scented with morpholine or an alternate chemical. Tracks show the responses of *A* imprinted salmon when morpholine was present, *B* imprinted salmon when morpholine was absent, *C* imprinted salmon when a chemical other than morpholine was present (e.g., PEA) and *D* non-imprinted salmon when morpholine was present. *Dots* along the track path represent 15-min intervals

as an "odor barrier" through which the fish had to swim. The morpholine concentration in this area was approximately 5.7×10^{-10} M. Water currents were measured with drogues to determine how long the chemical remained there. Before tracking, the fish were held for 3 to 17 days in tanks containing Lake Michigan water to reduce the possibility that they might react to morpholine only because they had been recently exposed to the odor.

We tracked 56 fish through the test area. Most of them remained near the release point for about 1 h before moving. Thereafter, they usually travelled at a constant speed and without changing direction on a route that paralleled the shoreline, within 50 m of shore. In all cases (20 tracks) when morpholine was present in the test area, morpholine-exposed fish stopped their migration and remained in the scented area from 1 to 4 h. Their stay correlated roughly with the time it took for water currents to dissipate the chemical.

When morpholine was not present in the test area, morpholine-exposed fish (14 tracks) moved through without stopping. Five different trackers observed at least two pairs of responses. These results demonstrate that morpholine invariably arrested the progress of morpholine-exposed fish. However, it was also possible that the fish reacted to morpholine simply because it was not the same as Lake Michigan water, i.e., because it was an unusual scent not normally encountered along the shore. If so, their behavior was not necessarily associated with a long-term memory of morpholine.

To test for this possibility we tracked 13 unexposed (control) fish through the test area when morpholine was present and all 13 moved through it without stopping. In addition, morpholine-imprinted fish were tracked through the test area when it was scented with a different chemical; none of these fish stopped their migration. In these tests the chemicals were phenethyl alcohol (7 tracks) and N-β-hydroxyethyl-morpholine (2 tracks). Thus it appears that morpholine-exposed fish were not reacting to morpholine as a unique shoreline odor. Rather their reaction seems to be associated with olfactory imprinting and long-term olfactory memory.

2.2.4 Electroencephalography

Electrophysiological experiments were conducted along with the behavioral studies at Oak Creek in 1972 and 1973 (Cooper and Hasler 1973, 1974, 1976). EEG responses to morpholine and other water samples were tested in 50 morpholine-exposed and 40 unexposed (control) fish captured in Oak Creek. Although there was a significant difference in the amplitude of the EEG signals to morpholine in distilled water for morpholine-exposed compared to control fish, there were no significant differences when other substances, such as phenethyl alcohol and Lake Michigan water were tested. Additionally Oak Creek water samples were collected above and below the dam where morpholine was being metered into the stream; hence fish could be tested with Oak Creek water containing or not containing the chemical. Morpholine-exposed fish responded to Oak Creek water containing morpholine but did not respond to Oak Creek water alone (Fig. 2.5). These electrophysiological experiments suggest that early exposure of juvenile salmon to morpholine significantly influences the subsequent responses of the olfactory bulb of sexually mature adults, a result that correlated well with the behavioral studies.

2.2.5 Electrocardiography: Conditioning the Heart Rate of Salmon as an Indicator of Their Response to Synthetic Chemicals

In another set of experiments, Hirsch (1977) used cardiac conditioning techniques to determine if olfaction is the sensory modality used by salmon in the detection of synthetic chemicals. In his experiments the heart rate of coho salmon, enclosed in acrylic tubes supplied with a constant water flow,

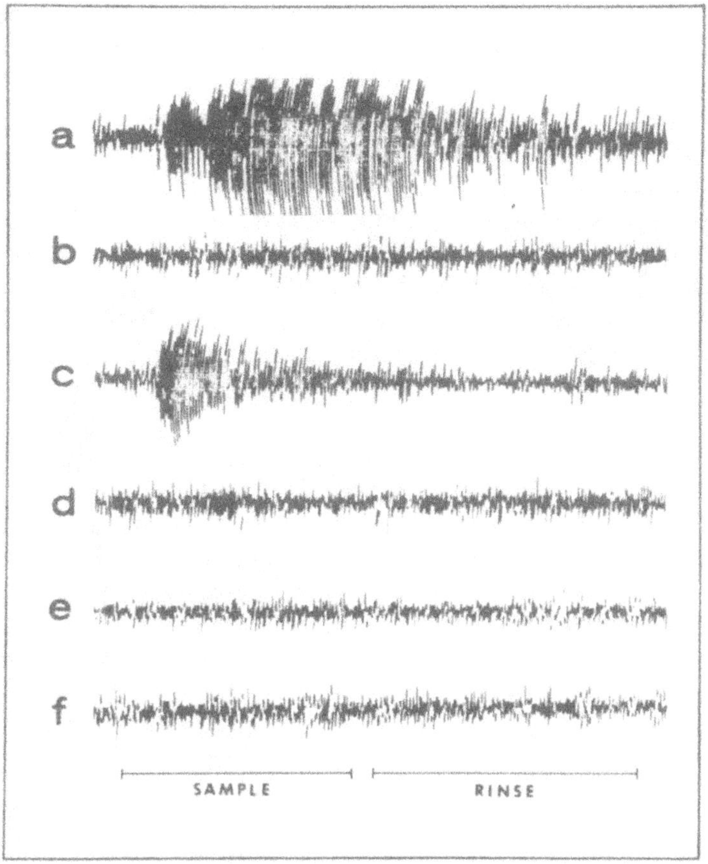

Fig. 2.5. EEG responses of (*a*) imprinted salmon to morpholine, (*b*) non-imprinted salmon to morpholine, (*c*) imprinted salmon to Oak Creek water containing morpholine, (*d*) imprinted salmon to Oak Creek water alone, (*e*) imprinted salmon to phenethyl alcohol (PEA), (*f*) imprinted salmon to N-β-hydroxyethyl-morpholine

could be conditioned to either morpholine or phenethyl alcohol by using electric shock. Odor cues of morpholine or phenethyl alcohol were presented, followed by a 3- to 5-V shock. Two groups of ten fish each were conditioned either to morpholine 5.7×10^{-8} M or to PEA 4.1×10^{-8} M. After 10 to 20 trials, the heart rate slowed down when the conditioning chemical was introduced into the testing apparatus before the shock was administered. The fish responded in this manner only to the chemical to which they had been conditioned. Presentation of other chemicals had no apparent effect on the fish.

Fish with plugged nostrils did not become conditioned to the presentation of odor and shock stimuli, though they did become conditioned after their nares were unplugged. Occluding the nasal chambers of conditioned fish eliminated their response to the odor. Presumably, if taste or a generalized chemical sensitivity were important in the detection of the synthetic chemicals, fish would still be able to respond to them since the chemicals were delivered through the water supply in which the fish were swimming. Fish conditioned to respond to morpholine did not respond to phenethyl alcohol, although they could be conditioned subsequently to phenethyl alcohol. This suggests that fish were able not only to detect the odor stimulis, but also to distinguish between odors.

More recent heart-rate experiments indicate that unconditioned adult salmon that were imprinted to morpholine or phenethyl alcohol as juveniles can still distinguish their imprinting chemical 18 months later (Hirsch personal communication).

William Sandoval (1980), working in Carl Schreck's laboratory at Oregon State University in Corvallis, has used both behavioral and cardiac-conditioning techniques for demonstrating recognition and discrimination of the odors of morpholine and phenethyl alcohol by coho slamon. He found that coho salmon recognize a mixture of morpholine and phenethyl alcohol distinctly from either odor individually. Fish could also distinguish a single odor to which they had been conditioned in a mixture containing that odor. Hence, his evidence argues that fish learn the "distinctive fragrance of the entire bouquet of odors", and, once having learned, can differentiate the bouquet from other odors.

2.3 Natural Imprinting

Since our field studies were corroborated by the laboratory work, they provide strong evidence that coho salmon retain and use synthetic chemical information to achieve successful homing. It seems likely that salmon with this ability would use it with natural chemicals in the environment. Additional support for this conclusion comes from two studies that show that salmonids become imprinted to natural water. Jensen and Duncan (1971) transplanted presmolt coho salmon from Leavenworth Fish Hatchery on the Wenatchee River, Washington, to a spring-fed fish-holding facility approximately 100 km away on the Snake River (see Fig. 1.1). The fish were marked, held for 48 h until they began to smolt, and then released in the Snake River. During the spawning migration, marked fish were recovered near the spring-water discharge 0.8 km downstream from the release point. No fish were recovered at Leavenworth Hatchery. To determine whether the fish were actually homing to the water in which they had been held as smolts, water from the holding facility was pumped through a floating trap.

Table 2.6. Number of coho salmon captured by a trap at ice harbor dam in relation to the type of water presented (Jensen and Duncan 1971)

Water source	Date of capture	Number of salmon
Spring water (gravity flow)	Nov. 2	52
	3	208
	10	5
	11	2
	14	7
	15	18
	16	18
	17	26
Spring water (pumped)	Nov. 7	59
	9	4
	Total	399
River water (pumped)	Nov. 4	0
	5	0
	6	0
	8	0
	12	0
	13	0
	Total	0

As a control, river water was pumped through the trap on alternate days. No fish entered the trap when river water was used, but 399 fish were captured when spring water used (Table 2.6). Thus it seems clear that the fish formed a permanent attraction for spring water from the holding facility and were able to learn the characteristics of the water within 2 days.

In the second study, Stuart (cited in Scholz et al. 1978 b) marked a group of young brown trout in one branch of a forked stream that flowed into the Dunalastair Reservoir in Scotland. After the fish had migrated to the reservoir, all of the water from the home fork was diverted into a new channel, and the original channel was maintained by water from the second fork. During the spawning migration adult trout homed to the new channel in preference to the channel by which they had entered the reservoir. Stuart's observations clearly indicate that the fish homed to water originating from the home tributary, rather than to a specific home location, and are thus consistent with our own conclusions that it is a characteristic of the home water, specifically odor, that provides salmonid fishes with homing cues.

2.4 Mechanisms of Olfactory Orientation in Upstream Migration

The artificial imprinting experiments have demonstrated that adult salmon can be attracted into streams scented with the synthetic chemical that was present in their water at the time of smoltification. However, the studies did not delineate the mechanisms by which the odor serves as a cue for directing upstream migration and home-stream selection. Brett and Groot (1963), citing hydraulic reports, have pointed out that because (1) in natural river systems concentration gradients steep enough to elicit klino- or tropo-taxis do not occur except in localized areas and (2) water masses from different sources do not mix readily, so that discharge of turbid water from a tributary into a main river channel creates discontinuities, it seems unlikely that salmon could follow an olfactory corridor back to its source. Instead, Harden-Jones (1968) has argued that the imprinting odor serves as a sign stimulus for releasing a stereotyped behavior pattern – in this case swimming against a current. He argued that such a mechanism could function in the selective guidance and segregation of fish with different imprinting experiences if the presence of imprinting odor caused positive rheotaxis (upstream swimming against currents laden with home water), and the absence of imprinting odor caused negative rheotaxis. Thus, if a fish made the wrong choice at a stream junction, the imprinting odor would not be present and the fish would swim downstream until encountering it again. Several important observations have been made in connection with this notion.

2.4.1 Overshooting and Proving: Positive and Negative Rheotaxis

Tagging–recapture studies and ultrasonic tracking studies suggests that "overshooting" or bypassing the natal tributary, i.e., making a wrong choice at a stream junction, is a common occurrence in migrating coho and sockeye salmon. It is also known that these fish rectify their errors through "back-tracking", i.e., they eventually swim back downstream and reach their natal tributary (reviews by Hasler 1966, Foerster 1968, Hara 1970). We have personally observed overshooting in ultrasonic tracking studies and in conventional tagging studies on numerous occasions.

DeLacey et al. (1969) displaced chinook salmon that had returned to the University of Washington College of Fisheries pond to locations in the Lake Washington watershed upstream and downstream from that site. Over 70% of the fish displaced downstream and 60% of the fish displaced upstream returned to the pond, suggesting that the presence of the home odor evokes a positive rheotaxis and absence of the odor evokes a negative rheotaxis. In addition, fish deprived of their sense of smell were captured in a salt water gillnet in Puget Sound, having swum downstream from the release point. Thus it is evident that the salmon tend to move downstream in the absence of home-stream odor. Kleerekopfer (1969) and Dodson and Legget (1979), investigated the problem by placing salmonids in an optomotor tank, and confirmed that odors do elicit rheotropic responses in salmon.

2.4.2 Rheotropic Responses of Salmon to Tidal Currents

It might be expected that rheotropic behavior would be reflected in the response of salmon to tidal or seiche currents at the river mouth. Hasler (1966) and Kleerekoper (1969) postulated there would be a tendency for salmon to swim against ebb or downstream currents laden with home-stream odors and to swim locally (i.e., "milling") or drift with flood or upstream currents.

The response of adult salmon to water currents during the spawning migration was measured in two separate studies using ultrasonic tracking techniques (Scholz et al. 1972). In the first, the migratory movements of 29 chum salmon were studied in a 65-m deep salt-water bay, 7.0 km long (E–W axis) by 3.0 km wide, in northern Honshu, Japan (Fig. 2.6). Their home river was located at the head of the bay. Fish were captured in a trap near the river mouth, released at the entrance of the bay, and then tracked back to the home river. We concurrently measured water currents and mapped salinity and temperature within the bay to determine the flow pattern from the home river. At ebb tide, river water flowed out of the bay along the south shore in the upper 10 m, but if the wind blew from the South West, river water flowed out along the north shore. At flood tide there was no flow from the river. The tracks of the salmon showed: (1) They swam into the bay toward the home river only during ebb tides. During flood tides they made no progress into the bay. (2) If the river water moved out of the bay along the south side, they swam into the bay by that route (Fig. 2.6). When river water moved out along the north side, they swam toward the river along the north shore (Fig. 2.6). (3) Visual observations indicated that the fish were swimming in the upper 3 m, i.e., at a depth which contained water from the home river. We base this on the fact that we often saw our transmitter-equipped fish, identified by an external red floy tag, swimming just below the surface when we approached them to record a position fix.

In a second study involving a landlocked freshwater run to coho salmon in Lake Michigan, 11 salmon near the mouth of the home river and 26 individuals swimming within the home river were observed reacting to periodic seiche current reversals. Salmon approached the mouth of the home river on both "ebb" and "flood" seiche currents, but entered the river

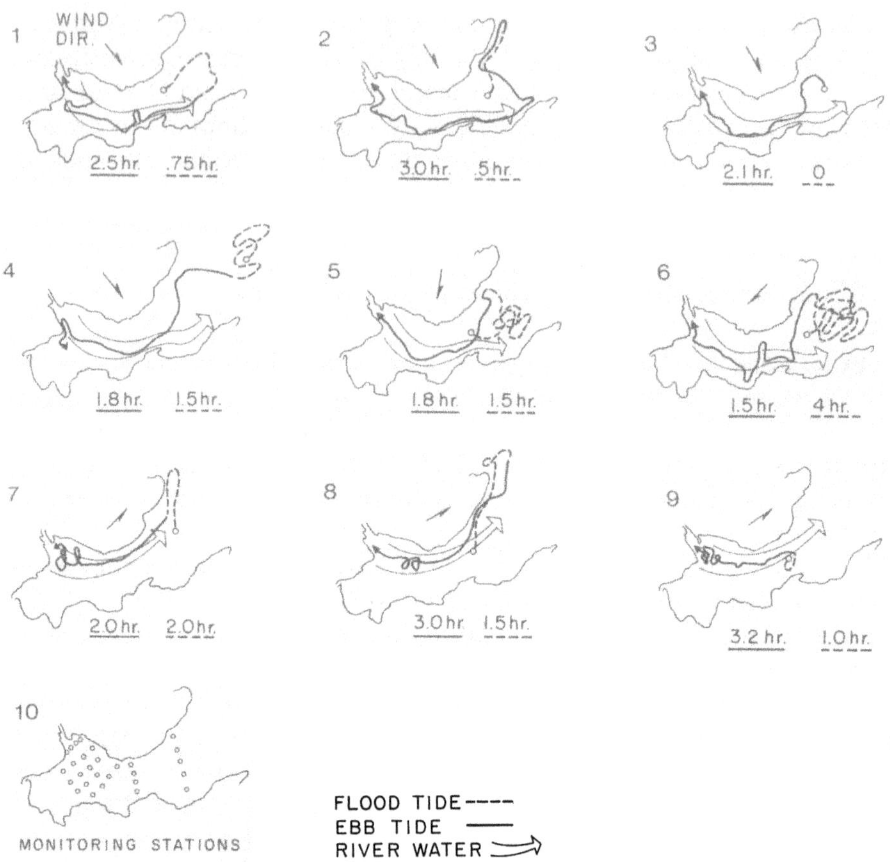

Fig. 2.6. Tracks of salmon in Otsuchi Bay made in 1969 and 1970. *1–6* Tracks on days when river water flowed out of the Bay along the south shore. *7–9* Tracks on days when river water flowed out of the Bay along the north shore. *10* The path of the river water was traced by measuring salinity. *Dots:* locations where salinity and tidal currents were monitored. *Solid lines:* portion of the track on the ebb tide; *dotted lines:* portion on the flood tide. Fish made progress into the Bay on the ebb and held position on the flood tide

only during ebb current conditions. Movements within the river below the home stream were almost exclusively upstream during ebb seiche currents but were generally localized during flood currents (Fig. 2.7). Dye-marking studies indicated that during flood currents water from the home tributary did not enter the main channel of the river. In addition, water from the home tributary flowed down the north side of the main river, and the fish traveled predominantly along that shore (Fig. 2.7). Four blinded fish displayed the same type of behavior as normal fish but five nose-plugged fish did not. The plugged fish drifted with water currents and eventually swam downstream to Lake Michigan.

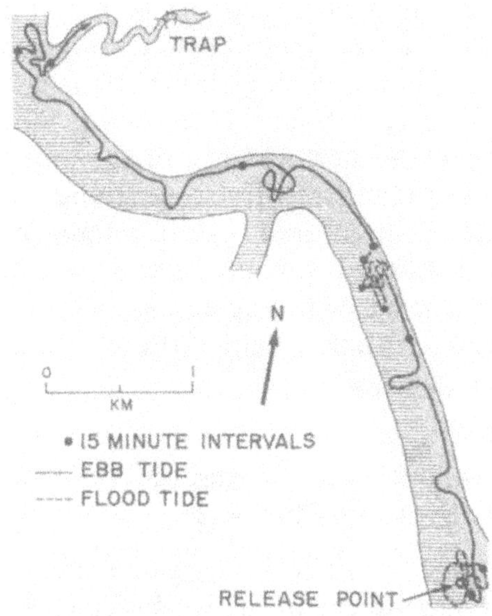

Fig. 2.7. Track of a salmon in the Ahnapee River near Algoma, Wisconsin. The track was conducted after a rainstorm. Discolored, muddy water from the home tributary was observed to enter the main channel and flow downstream along one shoreline (*stippling*). Fish traveled upstream predominantly along that shore. In addition, fish swam upstream on "ebb" seiche currents (*solid lines*) but swam locally on "flood" seiche currents (*dotted lines*)

Other tracking studies have also demonstrated a clear response by salmon to water currents. Madison et al. (1972) analyzed data from 18 sockeye salmon tracked in the ocean about 20 km off the coast of British Columbia. Two measures were used for detailed analysis of tracks: swimming speed and turning angle (i.e., change in course direction). The fish exhibited a diurnal rhythm of activity: swimming speed was greatest and course direction change least at mid-day (i.e., they traveled in straight lines for long distances). At night their swimming speed was reduced and the amount of turning increased (i.e., they milled around). There was a distinct change in behavior after the fish entered the estuary of their presumed home river. Their daily rhythm ceased and was replaced by responses to bidirectional current flow set up by tidal cycles. They swam against the ebb, and held position during the flood irrespective of the time of day.

Stasko (1975) tracked Atlantic salmon through the river-like Miramichi estuary on the Gulf of Saint Lawrence in New Brunswick. Nine fish were tracked spanning 71 flood and ebb tides. Fish that achieved overall upstream progress did so by drifting with flood tidal currents and stemming the ebb currents. Other investigators have observed the same type of behavior (Hallock et al. 1970, Legget 1973, Groot et al. 1975, reviewed by Stasko 1975, reviewed by Legget 1978). In addition, Banks (1969) and Arnold (1974) extensively reviewed hundreds of tagging–recaptures dealing with the response of salmon to currents and found that in most cases salmon swim upstream on ebb tide and hold position on flood.

In all of these examples, fish responded differentially to flood (by drifting, milling) and ebb (by directed upstream movement) currents. This suggests that their response is more than just a simple rheotaxis; instead, the fish are somehow able to distinguish between the flood and ebb. Improved orientation during ebb currents is consistent with the hypothesis that the presence of home-stream odors motivates the fish to swim against the current. Further evidence of the importance of odor is provided by the observations from tracking studies (e.g., Scholz et al. 1972) demonstrating that salmon, in an estuary, tend to migrate toward the home river on the side of the bay containing river water and are at the correct depth to be in it instead of taking equally likely alternative routes. Likewise, in a river, salmon traveled along the side that contained water from the home tributary.

2.4.3 Rheotropic Responses of Salmon Imprinted to Synthetic Chemicals

Johnsen (1978) investigated rheotropic responses in salmon imprinted to synthetic chemicals for his doctoral research. He imprinted coho salmon smolts to morpholine or phenethyl alcohol in a hatchery, stocked them in Lake Michigan, and captured them as adults in streams scented with the chemicals. The fish were then transported to and released in a different section of the river located upstream from the capture site. Morpholine was either introduced or not introduced upstream from that point. Johnsen expected that, if the imprinting odor acts as a sign stimulis for releasing positive rheotaxis, then morpholine fish would swim upstream if morpholine was present and downstream if absent. Phenethyl-alcohol fish would not be expected to move upstream in either case because their imprinting odor was never present in the study area. Morpholine-imprinted salmon migrated upstream when morpholine was present and downstream when it was absent. Phenethyl-alcohol-treated fish swam downstream in both cases. These results imply that the presence of the imprinting odor elicits positive rheotaxis, while absence of the odor evokes negative rheotaxis (Table 2.7).

Table 2.7. Results of tracks with morpholine present or absent from study area. (After Johnsen and Hasler 1980)

	Treatment group	No. tested	Up	Down	No movement
Morpholine present	MOR	18	14	2	2
	PEA	19	2	15	2
Morpholine absent	MOR	4		4	
	PEA	3		3	

Johnsen also introduced morpholine on either the right or left side of the stream, and correlated the routes of individual fish making upstream progress with the position of the odor trail. Flow patterns of the chemical were approximated by charting the distribution of a brilliant orange dye (Rhodamine B) introduced at the same point as the morpholine before and after each experiment. Dye introduced on one side of the channel did not extend beyond mid-stream (see Fig. 2.8). Movements of morpholine fish were confined to the right half of the river when morpholine was introduced on the right side, and the left half when morpholine was introduced on the left side (Fig. 2.8). If a fish swam out of the odor trail, it swam downstream until encountering the odor again.

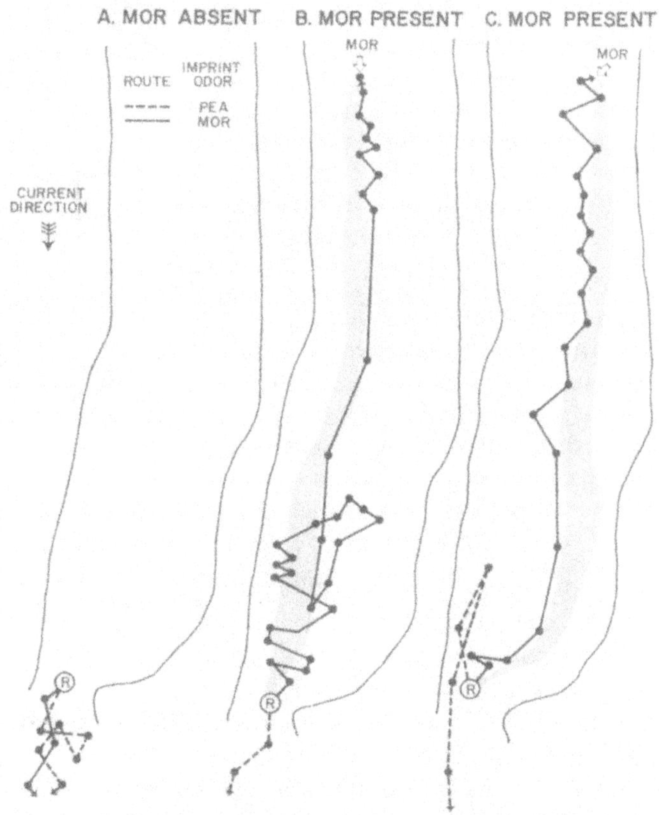

Fig. 2.8 A–C. Tracks of morpholine (MOR)- and phenethyl-alcohol (PEA)-imprinted fish with (*A*) morpholine absent, (*B*) morpholine present on the left side of the river, and (*C*) morpholine present on the right side of the river. When morpholine was present, morpholine-treated fish swam upstream; when it was absent, they swam downstream. Phenethyl-alcohol-treated fish swam downstream in both cases. In addition, morpholine-treated fish "tracked" the odor upstream, migrating upstream on the side of the river scented with morpholine along the edge of the odor trail. (After Johnsen 1978)

Another interesting aspect about the tracks is that, although the fish swam on the side of the river containing their imprinting odor, they traveled predominantly along the edge of the odor trail, passing in and out of it instead of traveling continuously within it. Hasler had previously argued that if salmon travel continuously within the odor trail, their olfactory sense would soon become adapted or "fatigued" to their imprinting odor so that they would fail to recognize it; whereas if salmon proceed in an interrupted (zig-zag) course, their olfactory sense would not become adapted and the scent would remain meaningful (Hasler 1966). Therefore, he predicted that salmon following an odor track would proceed in a criss-cross course along it, passing in and out of the odor-laden currents and responding to presence or absence of the odor as they progress upstream instead of remaining continuously within the odor track. Johnsen's results conform remarkably to this prediction.

2.5 Pheromones and Homing

A homing hypothesis that involves pheromones emitted by juvenile fish as a source for attracting adults has been advanced by several authors (reviewed by Horrall 1981 and Liley 1982). This idea was first suggested by G. H. Parker (cited in Chidester 1924) who thought "it is possible that a certain race of fish may give off emanations that differ chemically from those of other races; hence, the return of individual races to their homestream could be attributed to their power to sense the familiar emanation." In support of this idea White (1934) and Solomon (1973) reported that streams that had previously been barren of salmon became attractive to migrating adults shortly after juveniles were transplanted into them. White (1934), for example, planted fry in a stream that previously had never contained salmon. Later that year adults migrated up the river for the first time in recorded history, suggesting that "homing is influenced by population specific pheromones or metabolic products released by juveniles." The major drawback with this evidence is that no information was available which could definitely prove that the adults were genetically related to the juveniles. An equally likely alternative is that adults could have been attracted into the stream by some generalized conspecific odor. The problem, if the latter interpretation is correct, is that generalized olfactory attractants cannot account for the site specificity involved in homing. In this context, in both behavioral and electrophysiological tests, we have noticed that toward the end of the spawning season adults cease to respond to their imprinting odor, morpholine or phenethyl alcohol, and, instead, begin to respond strongly to the odors of other salmon (see Chap. 4 for details). A major point here is that the switch occurs after most of the adults have already migrated up the home stream and into the home tributary. Thus we feel that conspecific odors do not provide specific homing cues but instead may act as generalized attractants providing fish which have failed to home correctly with a mechanism for attracting them to sites with other spawning adults, thereby allowing for completion of their life cycle. This argument is developed more fully in Chapter 4.

In a more advanced formulation of the pheromone hypothesis, Nordeng (1971, 1977) proposed: (1) populations or races of salmon in different streams emit pheromones that serve to identify distinctly fish from that particular river, (2) the memory of this population-specific pheromone is inherited, i.e., stored in genetic memory, and (3) homing adults follow pheromone trails released by juveniles which reside in the stream, i.e., the juveniles provide a constant source of population odor. Nordeng's hypothesis is supported by a variety of behavioral and electrophysiological data which suggests that artic char, *Salvelinus alpinus*, are able to discriminate between different populations of their species on the basis of odor (Nordeng 1971, 1977, Døving et al. 1974, Selset and Døving 1980). Also, Nordeng found that mature males, kept isolated for 4 years, homed to a river that contained other members of their own population.

Even though evidence is accumulating that salmon may use pheromones as cues for homing, it is also clear from numerous successful transplant experiments (refer to Chap. 1) that homing involves imprinted as well as genetic memory. Moreover, Selset and Døving (1980) in discussing their work on migratory char state: "our results do not exclude olfactory imprinting, but make it likely that the possible imprinting would be to substances emanating from the fish and not from vegetation or minerals".

In our artificial imprinting work with coho salmon, rainbow trout, and brown trout, it is unlikely that fish homed to pheromones. In all of our experiments, there were no young coho salmon present in the scented streams at the time the adults were attracted to them. Since warm summer stream temperatures preclude survival of juvenile salmon in Lake Michigan tributaries, all production occurs in hatcheries separate from the Lake Michigan drainage system The fish are not stocked in the streams until just before smolt transformation and reside in them for only a brief time, usually from 2 to 4 weeks, before migrating to the lake. Although it is conceivable that the juveniles could impart their odor to the stream system by accidentally scraping off mucus on rocks, we feel this is an unlikely possibility in view of the short period of time the juveniles are actually in the stream. We can be even more definite about our work at Oak Creek. There are no records of, and local fish managers confirm that, no juvenile salmon were ever stocked in Oak Creek. Additionally, we conducted extensive electrofishing operations in Oak Creek from late winter through late spring and again from late summer through late fall every for 3 years and never captured a juvenile salmon. Despite the fact that the stream was barren, morpholine-imprinted fish were attracted to it when morpholine was present but not when morpholine was absent.

Another reason why we feel that pheromones could not have contributed to our results in any significant way is that we observed differential responses in the behavior of fish with different imprinting experiences. These fish were from the same spawning stock und randomly separated into groups for imprinting to different odors. According to the pheromone hypothesis, our different experimental groups should have displayed uniform behavior because they were related. Consequently, the fact that different experimental groups behaved distinctively would seem to rule out the possibility that our results could be explained by the pheromone hypothesis. Additionally, Sutterlin et al. (1982) found that Saint Croix River salmon, reared in a distant fish hatchery (150 km from the Saint Croix River) and released as smolts at a marine site 12 km from their parent streams, returned to the marine site after having spent 13–25 months at sea. A group of smolts chemically imprinted with morpholine in salt water had the highest rate of return. Their results are difficult to explain in terms of the proposed pheromone hypothesis because no juvenile salmon from the Saint Croix River were present at the salt-water-imprinting site, and yet the experimental fish homed there. The experimental fish homed preferentially to this site even though their parent stream, with a population of Atlantic salmon presumably emitting race-specific pheromones, was located only 12 km away. Sutterlin et al. (1982) concluded: "If such pheromones are operative, it would appear that their influence can be over-ridden by other directive factors."

Our data illustrate clearly that salmon are able to home in the absence of pheromones. However, in natural populations, genetic memory of pheromones, in addition to olfactory imprinting, may provide homing cues for adult fish. Available evidence indicates that there are both genetic and imprinted components to migration and home-stream selection in salmonids (Bams 1976, refer to Sect. 1.5.3). Possibly, salmon may have redundancy mechanisms for orientation as has been reported in birds. Also, although specific genetic programs are important for adapting salmon to local environments, an open-ended genetic program, such as the capacity for imprinting (learning) information about the natal tributary, has the effect of enriching their behaviored repertoire (Mayr 1974, Horrall 1981). Mayr (1974) has argued that this is a great advantage in terms of "being able to store more complex objects of behavior than the genetic program."

Part II
Hormonal Regulation of Smolt Transformation and Olfactory Imprinting in Salmon

Allan T. Scholz, *Robert W. Goy*, and *Arthur D. Hasler*

Chapter 3

Factors Influencing Smolt Transformation: Effects of Seasonal Fluctuations in Hormone Levels on Transitions in Morphology, Physiology, and Behavior

Olfactory imprinting in salmonids occurs during a particular phase of development – the smolt stage – when juveniles undergo marked transitions in morphology, physiology, and behavior just prior to their seaward migration. The adults use this olfactory information as a cue for homing during the spawning migration. Evidence for this stems from studies in which coho salmon exposed to synthetic chemicals, either morpholine or phenethyl alcohol, during the smolt stage were attracted to rivers scented with the chemicals as adults (Scholz et al. 1976, reviewed by Hasler et al. 1978; see Chap. 2). The two stages of the salmon life history of particular interest in imprinting studies are (1) the smolt stage when olfactory imprinting to the home-stream odor (or synthetic chemical) occurs, and (2) the homing migration when adults "recall" the home-stream odor (or synthetic chemical) even though they had not been exposed to it during their period of residence in the ocean or lake.

Smolt transformation is a complex process in which the physiological, morphological, and behavioral transitions are external indicators that the critical period for imprinting has occurred. The temporal coincidence between smolt transformation and olfactory imprinting has caused us to speculate that the imprinting process could be activated by the same factors that regulate smolt transformation. A considerable volume of work thereon has been accomplished (reviewed by Hoar 1976, Thorpe and Morgan 1978, Folmar and Dickhoff 1980, Scholz 1980).

This chapter is a comprehensive review of smolt transformation, including descriptions of the morphological, physiological, and behavioral transitions and control of this metamorphic process. The investigations described here provided the basis and rationale for our experiments on hormonal regulation of smolt transformation and olfactory imprinting which will be presented in Chapter 5.

3.1 Morphological, Physiological, and Behavioral Transitions

In a recent review Hoar (1976) defined smolts as small migrant or nomadic salmons, noting that the term had been used for more than 500 years in connection with the seaward-migrant stage of the salmon. Parr have mottled yellowish-green to brown coloration, osmoregulatory systems adapted to freshwater, and are solitary and territorial. Modifications in morphology, physiology, and behavior during smolt transformation in salmon include: (1) development of silvery coloration, (2) increase in salinity tolerance and osmoregulatory capability, and (3) loss of territorial behavior and increased migratory activity.

3.1.1 Development of Silvery Coloration

In fingerlings the body coloration is dark green or yellowish-brown with parr marks present. Parr marks are black vertical bars along the side of the body formed by aggregations of melanophores (Robertson 1948). These pigment cells are in a fully expanded state in parr. During the smolt stage the parr marks disappear and the flanks turn silver, a result of subcutaneous deposition of guanine crystals (Robertson 1948, 1949, Johnston and Eales 1967, 1968, 1970). The crystals overlap and interlock, forming a surface that acts like a reflecting mirror, hence producing a silver color if illuminated and viewed from above. The guanine is produced by iridocytes, i.e., goblet cells embedded in the integument. Also, the melanophores in smolts are either in a state of partial contraction or of actual disintegration (Robertson 1948).

Wagner (1970) and Chrisp and Bjorn (1978) have developed categories for classifying the smolt status of salmon based on the degree of silvering. Their classification scheme is as follows:

Parr: No apparent silvering. Body color dark green to yellowish-brown. Parr marks dark and clearly delineated.

Intermediate: Moderate degree of silvering. Parr marks either light in color and distinguishable under a thin purine layer (in early stages) or indistinct (in later stages). The disappearance of parr marks occurred first from below the lateral line (transition phase 1), afterwards from the anterior and posterior (transition phase 2) and, finally from the central part of the body (transition phase 3).

Smolt: Silver. Parr marks absent. Fish at this stage were observed from late April to late May. The silvering of the body surface became so condensed that the parr marks were difficult to recognize. Also at this time, the backs of the fish became iridescent green or bluish-green.

Speculation about the development of silvery coloration in smolts has centered around the notion that in both fresh- and saltwater the type of coloration pattern affords protective camouflage in that particular environment

(Robertson 1948, Hoar 1976). In the river, the alternating light–dark color pattern in parr serves to help the fish blend in with the stream-bank vegetation. In the ocean, the silvery hue reflects the surface so the smolt would blend in with the surface and be less visible to predators below them in the water column.

3.1.2 Downstream Migratory Activity

Parr are solitary and display aggressive behavior in defending their feeding territory. They orient upstream and feed on aquatic insects that drift downstream with the water current. Smolts cease territorial behavior, form aggregations and begin to migrate downstream, traveling predominantly at night (Kalleberg 1958, reviewed by Hoar 1958, 1976, Thorpe and Morgan 1978).

The mechanism of the seaward migration of salmon smolts has been the subject of much study, speculation, and argument. One faction holds that the movement downstream is strictly a passive displacement of the fish by the water current (Huntsman 1952, Hoar 1953, reviewed by Arnold 1974). Proponents believe that the physiological mechanisms by which this impaired ability of the fish to hold position (reviewed by Thorpe and Morgan 1978) are achieved include:

1. osmotic stress – the fish become osmotically stressed as osmoregulatory transitions that preadapt the fish for tolerating sea water occur;
2. altered metabolism resulting in stressful conditions;
3. inability of the fish to maintain a visual position fix during evening twilight; and
4. increased buoyancy of smolts.

All of these modifications apparently result from alterations in biochemical pathways, which are, in turn, regulated by hormones. There is evidence for each of these ideas.

Respirometer measurements of swimming performance demonstrate that Atlantic salmon smolts increase ventilation and have poorer swimming performance with increasing current flow than parr, indicating that the fish may be experiencing stress (McCleave and Stred 1975, Thorpe and Morgan 1978).

Emigrating coho, pink, chum, and sockeye salmon display strong negative phototaxis, remaining inactive near the bottom in deep water during the day and rising to the surface at night (Hoar 1951, 1953, 1954, 1958, 1976, Ali and Hoar 1959). Hoar suggests that while the smolts remain in visual contact with the bottom they are able to perceive cues which provide them with positional information to orient into the current and maintain their station, whereas fish in midwater would lose these signals and be transported downstream. In contrast, coho parr seem to exhibit this photonegative response and remain near the bottom during day and night. Ali and Hoar (1959) have speculated that parr and smolts behave differently toward light because the composition of visual pigments in their retinas is different. Retinas of salt- and freshwater fishes contain different light-sensitive pigments – purple-colored rhodopsin in the former and rose-colored porphyropsin in the latter. Each pigment is sensitive to different wavelengths and is adapted to the light conditions of their respective environments. Retinae of euryhaline salmonids contain both types of visual pigment: those from freshwater-adapted parr yield more porphyropsin, while those of seaward-migrating smolts yield more rhodopsin. Porphyropsin is more sensitive in conditions of low light because it is sensitive to wavelengths that penetrate under conditions of turbidity; hence, parr might be sensitive to low light levels at night and be able to obtain visual cues to orient into the current while smolts would not.

Saunders (1965) found that smolts are more buoyant than parr, which would further increase their probability of displacement. The increased positive buoyancy in smolts resulted from adjustment of swim-bladder volume. Saunders reasoned that negative buoyancy would be advantageous for parr maintaining a feeding station in swiftly flowing portions of streams.

Field observations also suggest that the downstream migration occurs as a result of passive drift. Foerster (1929) and Barnaby (1944) observed that sockeye smolts in natural rivers drifted downstream tail first and actually swam against the current when negotiating areas with fast current; even so they were swept downstream. Ultrasonic tracking studies (Fried et al. 1978, LaBar et al. 1978, McCleave 1978, Solomon 1978, Tytler et al. 1978) of Atlantic salmon in rivers and estuaries suggest that the rate of downstream movement of smolts is generally much slower than the flow of the river, which suggests that downstream movement is intermittent and that the fish actually spend most of their time not migrating.

Critics of the passive-drift concept have argued that the downstream migration represents a volitional, directed movement and point out that in many cases in which both the velocity of currents and distance of smolt movement per unit time could be measured in the field, smolts traveled farther than expected by passive displacement alone (Johnson and Groot 1963, reviewed by Hartman et al. 1967; McCart 1967, Chrisp and Bjorn 1978). In addition, proponents point to the behavior of smolts in hatchery raceways. Anyone who has had the opportunity to witness smolts crowding the downstream end, pressing against the screens, realizes that their migration represents an active movement. In this connection, we have made some observations by releasing smolts at the upper end of a 50-m-long raceway and recording the rate at which they swam to the downstream end. We also measured water-current velocity with an electric-generator type of current meter. Velocity was 17.2 cm s^{-1} while smolts swam on the average 27.4 cm s^{-1} (number of observations = 25).

Groot (1965) reported results of tagging studies that indicated that the downstream migration of sockeye salmon through an intricate lake system was a well-oriented movement on a direct route, i.e., the fish migrated downstream more rapidly than the current through the lake and were not caught in eddies as would be expected if the fish were drifting passively. Ultrasonic tracking studies of smolts also lend support to this argument (Fried et al. 1978, LaBar et al. 1978).

Hoar (1953) also found that sockeye smolts held in an experimental tank swam vigorously and rapidly with the current. Stasko et al. (1973) report that Atlantic-salmon smolts swim actively downstream, display increased leaping behavior compared with parr, and are highly excitable and active when trapped.

Advocates of the directed-movement hypothesis believe that active downstream migration could result from activation of motor patterns generated by specific centers of neuronal networks in the central nervous system. Electrical stimulation of the certain portions of the telencephalon evokes a stereotyped motor response that resembles swimming activity, while stimulation of certain other regions evokes aggression. Treatment of these regions with certain hormones often mimics electrical stimulation. Hormonal stimulation of the telencephalon could play a regulatory role in activating motor networks involved in downstream swimming. Hoar (1939) and Baggerman (1960a, b) reported that in Atlantic or sockeye salmon thyroid gland activity increased prior to the onset of the migration, remained high throughout the migration period, and decreased before the termination of the migration. This suggests that "increased thyroid activity may be one of the factors inducing migratory disposition, whereas decreased thyroid activity may be a factor involved in terminating the migration season" (Baggerman 1960b). Also, Godin et al. (1974) have shown that pre-smolt Atlantic salmon treated with 6.43×10^{-11} M thyroxine, at a time when the thyroid gland is expected to be relatively inactive, showed reductions in upstream orientation and aggressive behavior and increased downstream migratory activity in comparison with controls injected with solvent alone. However, no studies have actually been conducted that show that the hormones bind to pattern-generator centers in the central nervous system at the time of smoltification.

A third line of contention is that the downstream migration is largely a result of biological interactions. Proponents cite evidence that downstream migration of Atlantic salmon in two rivers in the Soviet Union appeared to result from defensive behavior in response to predation by northern pike (*Esox lucius*); the dynamics and timing of the movement of salmon was related to the predatory-activity pattern of the pike (Bakshtansky et al. 1976a, b, Manteifel et al. 1978). In another case in which parr of coho salmon were stocked in a chinook-salmon stream, social interactions between the two species resulted in the displacement of the native chinooks (Stein et al. 1972). Chapman (1962) has speculated on intraspecific aggression in juvenile coho salmon as a cause of emigration.

We do not believe that biological interactions are a major factor in initiating seaward migration of salmon smolts for several reasons: (1) the movements described were usually not the mass exodus of fish that is normally attributed to the smolt transformation; (2) even in cases in which large numbers of fish were displaced no follow-through study was made to determine if the fish actually left the river for the sea, or if they simply remained in the river farther downstream; (3) most importantly, many of the fish in these studies left or were forced downstream at times that were completely inappropriate for smolt transformation.

Instead, we incline toward a view that incorporates the models of passive drift and directed movement. In some situations it is clear that the downstream migration is passive, while in others it is equally obvious that the downstream migration represents directed movements. Moreover, in many cases smolts from one location display both types of behavior. In slow to moderate current velocities, they orient downstream and swim faster than the current, whereas in turbulent water they orient upstream and drift downstream tail first (Stasko et al. 1973). Also, both models propose physiological mechanisms for the transition to the migratory state. Although the details of the activation mechanisms are different in each case, the overall scheme – i.e., hormonal mediation of biochemical pathways – is similar in both models.

Downstream migration is principally nocturnal in Atlantic, sockeye, and coho smolts. This has been measured by: (1) direct observation (Hoar 1953, 1954, 1958), (2) monitoring traps at different times during the day and night and enumerating the downstream migrants (Johnson and Groot 1963, Groot 1965, Munro unpublished, Hartman et al. 1967, reviewed by Forster 1968, Osterdahl 1969, Chrisp and Bjorn 1978, Thorpe and Morgan 1978); (3) acoustic survey estimates of smolt activity at different times of day and night (Johnson and Groot 1963); and (4) ultrasonic tracking (Fried et al. 1978, LaBar et al. 1978, McCleave 1978, Solomon 1978, Tytler et al. 1978). As an example of the type of information available on the timing of the downstream migration we will report the work of Hartman et al. (1967). At 2-h intervals from late April to late May they monitored traps that completely blocked the outlets in several sockeye-nursery lakes. In all cases, smolts migrated nocturnally, beginning at sunset with most intense activity occurring 3–4 h after sunset, i.e., 96% of the smolts migrated between 2100 and 0300 h. Since this pattern of activity was inferred from data on capture of smolts in traps, it has been suggested that the recorded periodicity might reflect the ability of the fish to avoid capture visually during the day rather than a genuinely entrained pattern of downstream movement. However, recent ultrasonic-tracking experiments, which were designed to reveal the time patterns of movements of unrestricted smolts, confirmed the pattern of nocturnal migration.

The downstream migration of most coho salmon lasts for 4–6 weeks, beginning in late April and continuing into early June (Hoar 1976). For other species, the migratory season is of approximately the same length, but the time of year may be shifted backward or forward by several weeks (Hartman et al. 1967, Thorpe and Morgan 1978). Hartman et al. (1967) calculated the cumulative percentage of the total population of sockeye smolts that display the pattern described above for coho:

Date		Interval	Cumulative Total
Apr.	25	End of 1st week	< 1%
May	2	2nd	5%
	9	3rd	40%
	16	4th	90%
	23	5th	99%
	30	6th	> 99%
June	6	7th	100%

Thus, a distinct maximum in migratory activity occurred with about 85% –90% of the smolts emigrating over a 2-week period from May 9th to 23rd.

Another facet of the smolt migration of interest is the relationship between the development of downstream migration activity and other physiological transitions. Although the "seaward" migration generally coincides with the development of silvery coloration, this is not always the case. In some situations distinct silvering may begin to occur several months before the downstream migration commences. In others, silvering may not take place until after the initiation of the migration. While this type of information is not recorded in the literature, we base our statements on discussions with other workers and our own casual observations. For example, we have personally observed distinct silvering in some groups of coho salmon in March. The fish were in the early-to-late intermediate stages described earlier, but showed no sign of downstream migratory activity. In another instance, we observed coho smolts migrating in the Columbia River system in eastern Washington. The fish had already migrated at least 200–300 km downstream since fish collected at a National Marine Fisheries Station at Little Goose Dam on the Snake River bore fin clips from Dworshak National Fish Hatchery on the Clearwater River located about 300 km upstream from the collection facility. These fish still retained distinct parr marks. Likewise, salinity tolerance can increase before or after the start of the downstream migration (reviewed by Chrisp and Bjorn 1978; Doug Weber, National Marine Fisheries Service Biological Laboratory, Seattle, Washington, personal communication).

In the groups of coho salmon from Lake Michigan we have used for our own work, development of downstream migratory activity, silvery coloration and increased salinity tolerance normally occur at approximately the same time. Our view, which we will develop in more detail later in the present chapter and in the following chapters, is that the various physiological and behavioral transitions are regulated independently by different hormones. The transitions generally occur at approximately the same time because the hormones involved seem to be responsive to fluctuations in daylength. However, the hormones governing some of the morphological, physiological and behavioral processes may also be responsive to fluctuations in other environmental factors (e.g., water temperature), and therefore be out of synchronization with the other transitions. In situations in which smolts must travel a long distance downstream the absolute synchronization of salinity tolerance or coloration with downstream migratory activity might not be critical.

3.1.3 Salinity Tolerance and Osmoregulation

The development of salinity tolerance and osmoregulatory capability is another important feature of smolt transformation. Three types of test are performed to assess salt-water adaptation in salmonids: salinity tolerance and preference, osmoregulatory capability, and gill Na^+/K^+ ATPase activity.

In studies of salinity tolerance and osmoregulation rainbow trout, coho salmon, and Atlantic salmon parr did not survive transfer to salt water whereas smolts did. Upon transfer of pre-smolts from freshwater to high salinities, blood osmotic concentration steadily increased until death occurred. In smolts transferred from fresh to salt water, the initial increase in blood osmotic concentration was followed by a subsequent decline which leveled off at approximately the same concentration as the fish had maintained in freshwater, showing that the fish were able to osmoregulate.

In the salinity-tolerance experiments, fish were tested at different developmental stages from alevin to smolt by direct transfer from freshwater to 10, 15, 20, 25, 30, or 35‰ (respectively 310, 450, 590, 730, 870, or 1,000 mOsm/liter) salt water and calculating percent survival after 4 days or LD_{50} – number of hours to death for 50% of the population. Similar investigations have been conducted with coho and Atlantic salmon and brown

Table 3.1. Comparison of salinity tolerance in salmon with different life histories. Chum and pink migrate to the ocean as fry and can tolerate high salinities immediately after emergence. Coho and rainbow spend 1.5 years in freshwater. Their fry stage is not tolerant of high salinities. Salinity tolerance develops during the smoltification process. (After Weisbart 1968)

Species	LD_{50} (h) in			
	Alevin	Fry	Fingerling (parr)	Smolt
Pink	70	>336	–	–
Chum	51	>336	–	–
Coho	<24	< 24	<24	>336
Sockeye	<24	< 24	<24	>336

and rainbow trout (Huntsman and Hoar 1939, Black 1951, Gorden 1959, Koch et al. 1959, Baggerman 1960b, Parry 1960, Houston and Threadgold 1963, Conte and Wagner 1965, Conte et al. 1966, Weisbart 1968, Wagner et al. 1969, Otto and McInerney 1970, Wagner 1974a, Farmer and Ritter 1978). Weisbart (1968) noted marked changes in resistance to salt water in alevins and recently emerged fry of chum and pink salmon, species that migrate to the ocean after emergence, but not in these stages of coho and sockeye salmon, species that do not migrate after emergence (Table 3.1). In coho and sockeye, resistance to salt water develops during the smolt transformation at the time they migrate to the ocean, i.e., at the onset of the seaward migration all four species show a decided preference for salt water.

In studies of osmoregulatory capability salmon were tested at regular intervals from parr to smolt by transfer from fresh to salt water at salinities ranging from freshwater to 35‰ for various time periods. Osmotic concentrations of blood serum and environmental water were determined by a vapor-pressure osmometer or by freezing point depression so that osmoregulation curves could be constructed with osmotic concentration of blood plotted against the concentration of environmental water. These studies show that smolts of Atlantic or coho salmon placed in salt water are able to regulate blood-serum concentration while parr are not (Huntsman and Hoar 1939, Houston 1960, 1964, Parry 1960, Potts 1970, Boeuf et al. 1978).

3.1.4 Osmoregulatory Correctional Mechanisms in Smolts: Gill Na^+/K^+-ATPase Activity

The correctional mechanisms underlying the development of salinity tolerance in salmonids have been extensively studied. The gills of salmon contain Na^+/K^+ ATPase which functions as an outwardly directed sodium-ion "pump". Energy from ATP is required. Sodium binds to the internal and potassium to the external surface of a protein carrier embedded in the gill membrane. As the ATP is split the energy released causes the carrier to "flip-flop" in the membrane, resulting in the movement of sodium out and potassium in. The important features about this system are: (1) since energy is used, sodium and potassium ions can be transferred out and in, respectively, against their concentration gradients, and (2) after entering the gill epithelial cell potassium flows back out along its concentration gradient, while sodium cannot re-enter because of the selective permeability characteristics of the membrane. Hence, the steady-state equilibrium is maintained far away from thermodynamic equilibrium. The activity of Na^+/K^+ ATPase is inhibited by a chemical agent, ouabain, which blocks the potassium-binding site. Sensitivity to ouabain is the criterion used to separate the activity of Na^+/K^+ ATPase from other ATPases which could function, for example, in inwardly directed ion transport.

Several investigators have demonstrated that gill Na^+/K^+-ATPase activity increases during the process of smolt transformation in several species of salmon. The procedure and conditions for optimizing the assay were reported by Maetz (1971), Johnson et al. (1977) and Ewing and Johnson (1978).

Zaugg and McLain (1972) observed that Na^+/K^+-ATPase activity in gills of coho salmon increases from 12 μmol ATP hydrolyzed mg^{-1} protein h^{-1} in January to a peak of 24 μmol ATP hydrolyzed mg^{-1} protein h^{-1} in May. The activity was constant from January to mid-April, when it began to rise. The increase was correlated with increases in salinity tolerance, degree of silvering and tendency to migrate downstream. Similar observations have been recorded for chinook salmon, Atlantic salmon, and rainbow trout, although in some instances Na^+/K^+-ATPase activity and salinity tolerance increased several weeks before either development of silvery coloration or downstream migration (Zaugg and McLain 1969–1972, Adams et al. 1973, 1975, Giles and Vanstone 1976, Kerstetter and Keeler 1976, Lorz and McPherson 1976, Johnson et al. 1977, Boeuf et al. 1978, Chrisp and Bjorn 1978, Ewing and Johnson 1978, Lasserre et al. 1978). In other studies Na^+/K^+-ATPase activity and salinity tolerance did not increase until after smolts actually began migrating downstream (Chrisp and Bjorn 1978).

Ewing et al. (1979) raised chinook salmon for 1.5 years and found the background level of activity remained fairly constant (10–12 μmol P_i mg^{-1} protein h^{-1}) with peaks during spring (14–16 μmol P_i mg^{-1} protein h^{-1}) and fall (16 μmol P_i mg^{-1} protein h^{-1}) of the first year of life and a much higher peak (20 μmol P_i mg^{-1} protein h^{-1}) during the second spring at the time of smolt transformation.

3.1.5 Fish with Freshwater Life Cycles

One interesting aspect about the development of salinity tolerance and osmoregulatory mechanisms are cases in which salmon remain, or are retained, in freshwater after smolt transformation has occurred instead of making the migration to salt water. Osmoregulatory mechanisms of fish that spend their entire life in freshwater – either naturally, as in the case of the land-locked sockeye (McInerney 1964, Conte et al. 1966), or owing to artificial propagation, as in the case of the Great Lakes coho salmon – like anadromous fish, begin to adjust to salt water while in their home tributary but become readjusted to freshwater conditions when they enter the lake (Baggerman 1960, Conte and Wagner 1965, Conte et al. 1966, Wagner 1974 b, Scholz 1980).

3.2 Comparative Aspects of Smolt Transformation

Other species of salmonid fishes differ from coho salmon with respect to the details of their life cycles. Not all species remain in freshwater after emerging from the gravel. Some (Atlantic, rainbow, brown, fall chinook, sockeye), like the coho, establish residence in the stream, normally remain there for 1.5 years and have a distinct smolt stage. Others emigrate immediately after emergence (pink, chum) or after 3 to 6 months (spring chinook) and do not have a distinct smolt stage, i.e., do not exhibit the distinct silvering of species that undergo the smolting process. However, the parr marks of these fry are not as well developed as in fry of species that do smolt, and the fish do exhibit other typical smolt characters, e.g., they swim downstream and are able to tolerate salt water.

In an attempt to understand how time of emigration is controlled, fry and smolts of different species have been studied with respect to salinity preference, tolerance and osmoregulatory capability, body coloration, and downstream migratory behavior (Hoar 1951, 1953, 1954, 1958, 1976, Keenleyside and Hoar 1954, Kalleberg 1958, Parry 1960, Baggerman 1960 a, b, McInerney 1964; summarized in Table 3.2). In this context the term "migratory behavior" encompasses behavior like hiding, territoriality, and schooling, in addition to components of downstream movement such as positive or negative reaction to water currents, active downstream swimming as opposed to passive drift, and responses to light.

In migratory salmon, e.g., fry of chum and pink or smolts of sockeye and coho, schools of downstream migrants exhibit a photonegative reaction to light and remain stationary under cover or in deep water during the day; as light intensity falls the fish rise toward the surface and increase their

Table 3.2. Comparisons of morphological, physiological and behavioral transitions between fish that migrate downstream as fry or smolts. Note that these transitions are correlated with time of downstream migration, not chronological age

	w/Smolt stage					Inter-mediate	w/out Distinct smolt stage		
	Coho	Spring chinook	Atlantic	Rain-bow	Brown	Sockeye	Pink	Chum	Fall chinook
Normal age at downstream migration (MO)	18	18	18	18	18	18	4	4	6
Recently emerged fry									
Salinity									
Tolerance	Low	Low	Low	Low	Low	Low	High	High	Medium
Preference	FW	FW	FW	FW	FW	FW	SW	SW	SW?
Osmoregulatory capability	No	No	No	No	No	No	Yes	Yes	Yes
Coloration									
Silver	No	No	No	No	No	No	No[4]	No[4]	No[4]
Parr marks present	Yes	Yes	Yes	Yes	Yes	Yes	No	No	No
Migratory activity									
Hiding	Yes	Yes	Yes	Yes	Yes	No	No	No	No
Territorial (nipping)	Yes	Yes	Yes	Yes	Yes	No	No	No	No
Schooling	No	No	No	No	No	Yes	Yes	Yes	Yes
Positive rheotaxis	Yes	Yes	Yes	Yes	Yes	No/yes	No	No	No
Negative rheotaxis	No	No	No	No	No	Yes/no	Yes	Yes	Yes
Photonegative response exhibited	No	No	No	No	No	Yes/no	Yes	Yes	Yes

Smolts						
Salinity						
Tolerance	High	High	High	High	High	High
Preference	SW	SW	SW	SW	SW	SW
Osmoregulatory capability	Yes	Yes	Yes	Yes	Yes	Yes
Coloration						
Silver	Yes	Yes	Yes	Yes	Yes	Yes
Parr marks present	No	No	No	No	No	No
Migratory activity						
Hiding	No	No	No	No	No	No
Territorial (nipping)	No	No	No	No	No	No
Schooling	Yes	Yes	Yes	Yes	Yes	Yes
Positive rheotaxis	No	No	No	No	No	No
Negative rheotaxis	Yes	Yes	Yes	Yes	Yes	Yes
Photonegative response exhibited	Yes	Yes	Yes	Yes	Yes	Yes

Abbreviations: FW = freshwater; SW = Saltwater

activity, which results in displacement downstream. Non-migratory salmon such as fingerling coho, rainbow, or Atlantic salmon, remain quiet at night and rest on the bottom of the stream in contact with specific objects in their environment, thereby maintaining their position and not drifting downstream.

These studies showed that preferences or behavioral responses of the fry differ among species. Also the preference or behavioral response in a single species changes over the course of development. The main point here is that, while the fry of species that emigrate shortly after emergence exhibit distinctly different behavioral and physiological responses than those of fry of species that remain in freshwater, they do resemble the smolts of those species. Certain types of activity or preferences seem to be associated with downstream migration in both groups. Moreover, imprinting at the time the young fish migrate downstream has been reported for all species (reviewed by Ricker 1972; summarized in Chap. 2):

Pink: Pritchard (1938), Wickett (1958), Hunter (1959), Neave (1966), Bams (1976)
Chum: Hunter (1959), Sakano (1960), Sano (1966)
Chinook: Rich and Holmes (1929); Ebel (1970), Ellis (1970), Ebel et al. (1973), Park (1975), Slatick et al. (1975)
Sockeye: IPSFC (1947–1976), reviewed by Foerster (1968)
Atlantic: White (1936), Carlin (1955, 1968), Mills and Shackley (1971)
Rainbow: Larson and Ward (1954), Shapovalov and Taft (1954), Ebel et al. (1973), Slatick et al. (1975), Weber (1975)
Brown: Stuart (1959)

3.3 Evidence for Hormonal Regulation of Smolt Transformation

The salmon endocrine system undergoes major alternations at the time of smolt transformation and is involved in regulating the morphological, behavioral, and physiological transitions associated with the smoltification process (reviewed by Hoar 1976; Folmar and Dickhoff 1980; Scholz 1980).

Thyroid involvement in smolt transformation is indicated by four major lines of evidence:

1. Histology. In studies in which thyroid glands are collected sequentially from pre-smolt to smolt stage and examined for histological signs of activation, the results indicate that thyroid activity increases before smolt transformation occurs. In coho, sockeye, and Atlantic salmon and in rainbow and brown trout glandular differentiation involves a transition from the cuboidal type of follicular epithelium (quiescent state) to the

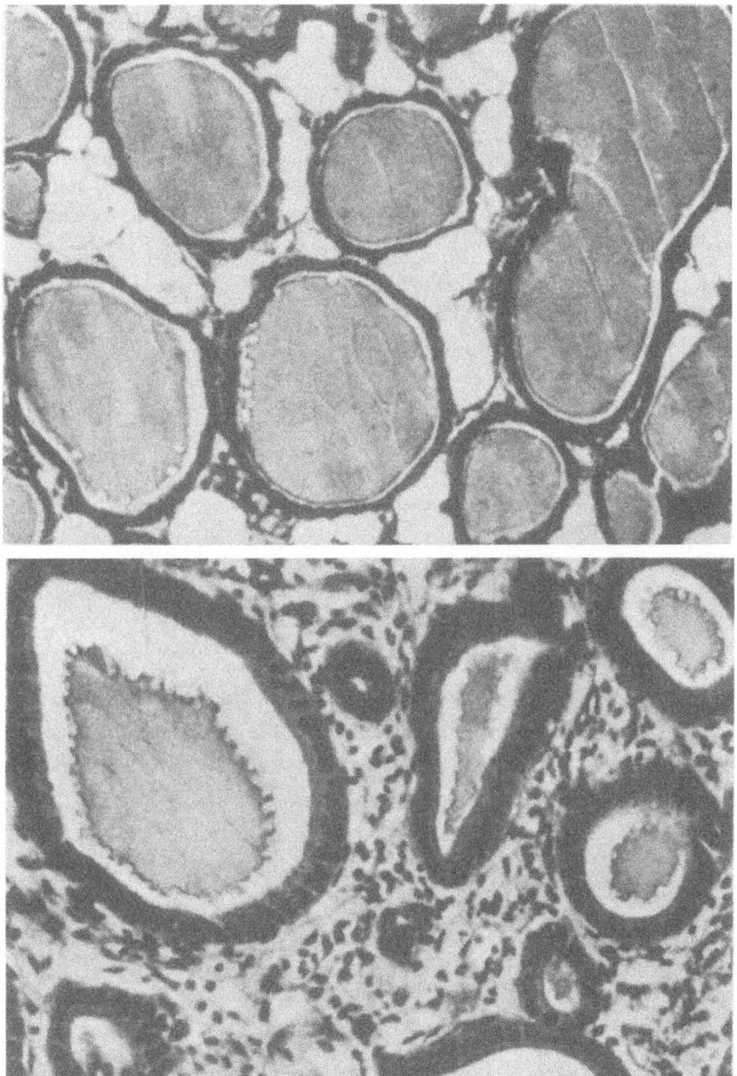

Fig. 3.1. Cross-section of thyroid follicles from a coho salmon parr (*top*) and smolt (*bottom*). In coho parr the follicular epithelium is of the low, cuboidal type indicative of a quiescent state. Also the lumen is filled with colloidal material (stored hormone). In the smolt the follicular epithelium is of the columnar type indicative of an active state, which is further evidenced by vacuolization and loss of colloid from the lumen

columnar type (active state), accompanied by vacualization and loss of colloid from the lumen (Fig. 3.1; Hoar 1939, Landgrebe 1941, Robertson 1948, Hoar and Bell 1950, Swift 1959, Baggerman 1960a, Eales 1963, 1965).

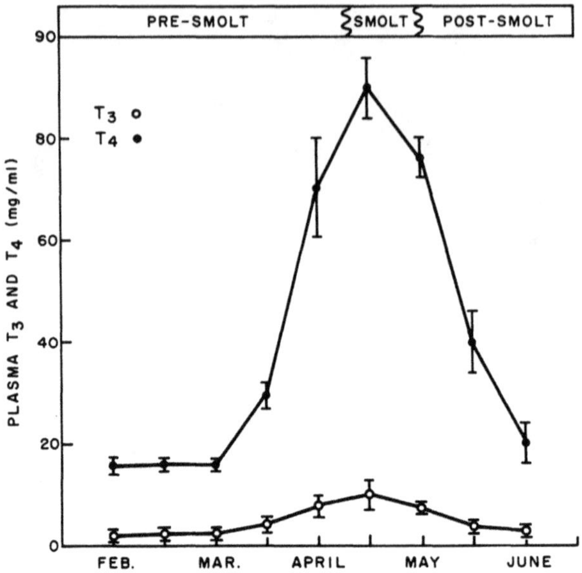

Fig. 3.2. Blood serum concentrations of triiodothyronine and thyroxine during the smolt transformation process. Note the thyroid hormone surge which occurs just prior to smolt transformation. (After Dickhoff et al. 1978)

2. Circulating levels of thyroid hormones. Direct measurements of circulating thyroid hormones have been made by radioimmunoassay (Dickoff et al. 1978, Scholz 1980). Blood serum from coho salmon was collected at 2-week intervals from winter through spring (i.e., pre-smolt through smolt stage) and was assayed for triiodothyronine and thyroxine. Triiodothyronine and thyroxine increased from approximately 2 to 10 ng ml^{-1} and 10 to 90 ng ml^{-1} serum, respectively, before smolt transformation, i.e., the surges of thyroid hormones, occurred at the beginning of April preceding smolt transformation at the end of April (Fig. 3.2). Radioimmunoassays of plasma from rainbow trout and chinook salmon also demonstrate that a maximum in circulating thyroxine levels occurs in the spring at the time of smoltification (Osborn et al. 1978, Folmar and Dickhoff 1979).
3. Administration of thyroid hormones. Injection of thyroid-stimulating hormone (TSH) or thyroid extract in a solvent into pre-smolts induces silvering and increases downstream migratory activity in coho salmon, rainbow trout, Atlantic salmon, and brown trout. Injection of solvent alone (control) does not produce these responses (Robertson 1949, Baggerman 1960 b, Godin et al. 1974). Administration of TSH or thyroid extract caused the loss of parr marks and distinct silvering in rainbow trout, brown trout, and Atlantic salmon (Landgrebe 1941, Robertson 1949). Administration of TSH and thyroid extract reduced upstream orientation and aggressive behavior, and increased downstream migratory activity and schooling in coho and Atlantic salmon (Hoar et al.

1955, Baggerman 1960 b, Godin et al. 1974). This suggests that thyroid hormone activation may produce the behavioral responses necessary for abandonment of stream residence.

4. Blocking normal thyroid function with antithyroid compounds or radiothyroidectomy. Radiothyroidectomy or administration of thiourea or thiouracil, which interfere with the biosynthesis and release of thyroid hormones, retards normal development and prevents normal smolt transformation in brown trout, rainbow trout, Atlantic salmon, and Pacific salmon (Dales and Hoar 1954, LaRoche and Leblond 1954).

Evidence for prolactin (PRL) effects on smolt transformation include:

1. Hypophysectomy and replacement therapy. Hypophysectomy in freshwater-adapted fish, including juvenile salmon, causes death. PRL injections enable hypophysectomized fish to survive, whereas other anterior pituitary hormones are ineffective. This suggests that PRL is an osmoregulatory hormone involved with maintenance of water balance in freshwater (reviewed by Johnson 1973, Bern 1975).
2. Histology. PRL cells in the anterior pituitary gland become inactive during smolt transformation in Atlantic salmon, rainbow trout, sockeye salmon, and coho salmon (Olivereau 1954, Zambrano et al. 1972, Leatherland and McKowen 1974).
3. Measurement of PRL. Electrophoretic or radioimmunoassay determinations of PRL content in Atlantic salmon indicate that circulating PRL declines during spring as smolt transformation occurs (Leatherland and McKeowen 1973, Komourdjian 1976).
4. Injection of PRL. PRL decreases Na^+/K^+-ATPase activity in the gills and blocks development of salinity tolerance (Ogawa 1974, Higgs et al. 1976, Clarke et al. 1977).

Evidence for influence of adrenocorticotrophic hormone (ACTH) or cortisol on smolt transformation includes:

1. Histology. Interrenal gland becomes active during smolt transformation (Olivereau 1954). So also do ACTH cells in anterior pituitary gland (Zambrano et al. 1972).
2. Measurement of corticosteroid hormones. Radioimmunoassays of plasma from coho salmon suggest that circulating levels of corticosteroid hormones increase during the smolt stage (Carl Schrek and Jennifer Specker, Oregon State University, personal communication, Scholz 1980)
3. Administration of ACTH or cortisol. Injection of ACTH increases tolerance to salt water, osmoregulatory capability, and Na^+/K^+-ATPase in gills of coho salmon (Scholz 1980).

Evidence for effects of growth hormone (GH) on smolt transformation includes:

1. Electrophoretic determinations of GH in the pituitary gland. These suggest that GH is released from the pituitary gland at a rapid rate during smolt transformation (Komourdjian 1976, Komourdjian et al. 1976 a, b).
2. Injection of GH into pre-smolts. Such injections increase salinity tolerance and salt water preference and promote differentiation of gill chloride cells (Clarke et al. 1977).

In summary, thyroid hormones seem to regulate downstream migratory activity and silvering, whereas PRL, GH and ACTH (cortisol) function as osmoregulatory hormones. Furthermore PRL and ACTH (cortisol) function antagonistically to each other (see excellent reviews by Bently 1971, Johnson 1973, Bern 1975):

1. Decreasing PRL and increasing ACTH (cortisol) during smolt transformation is associated with development of salt water tolerance of smolts.
2. PRL inhibits and cortisol activates Na^+/K^+-ATPase (outwardly directed sodium ion "pump"), in the gills.
3. Activation of this enzyme increases osmoregulatory capability, enabling the fish to tolerate salt water.

From the studies described in this section it is clear that seasonal fluctuations in the levels of several hormones are involved in regulating the morphological, physiological, and behavioral transitions that describe the smolt stage. In view of the temporal correlation between smolt transformation and olfactory imprinting we suspected that hormone fluctuations might also be involved in activating the imprinting process. A review of pertinent literature indicates that thyroid hormones and cortisol may influence olfactory sensitivity and memory function in salmon.

3.4 Hormonal Influences on Olfactory Sensitivity, Learning, and Memory in Salmonids

Fontaine (1975), from EEG studies dealing with odor detection in fish, concluded that spontaneous electroencephalograms from the olfactory bulb depend on the circulating levels of thyroxine. Additionally, thyroid-hormone injections modify electrical activity of the olfactory bulb (Oshima and Gorbman 1966 a, b, Oshima and Gorbman 1966 a, reviewed by Gorbman 1969). Moreover Oshima et al. (1972) have shown that the response of the olfactory bulb is modulated both by the forebrain and more psoterior parts

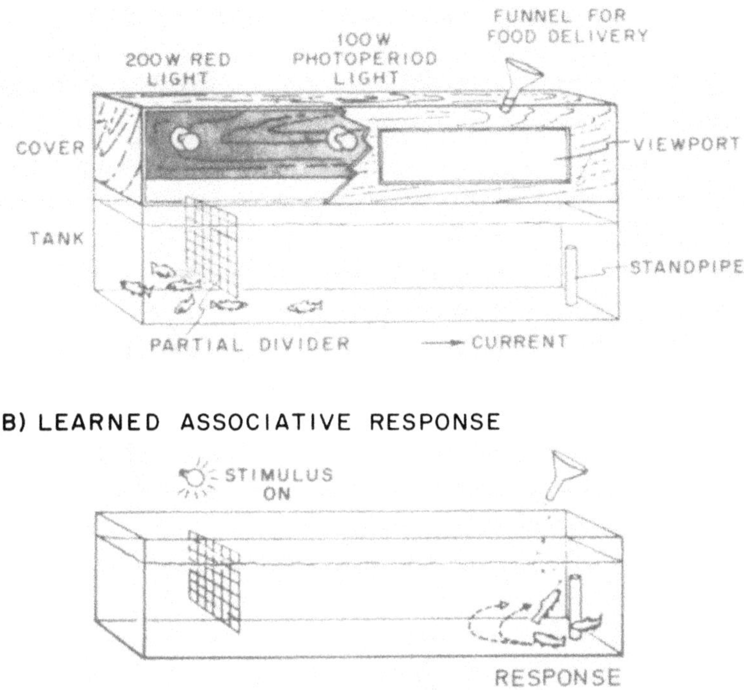

Fig. 3.3 A, B. Experimental chamber for experiments on memory acquisition and retention. Red-light flash (unconditioned stimulus) was coupled to food (unconditioned stimulus) delivery into the funnel. *Black arrow* denotes the typical response of fish to a flash of red light after it had been trained to associate the red-light flash with food

of the nervous system, and that both of these controls are separately sensitive to thyroxine.

TSH injection also facilitates memory in rainbow trout. Lindberg and Scholz (cited in Scholz 1980), in a preliminary study, found that pre-smolt fish trained to associate a red-light flash with food were able to learn faster and retain the information for longer periods if they had been treated with TSH instead of saline.

In this study a flash of red light (conditioned stimulus) was coupled with feeding (unconditioned stimulus). Movement of the fish toward the red light and then toward the feeding station or directly to the feeding station was scored as a positive response (Fig. 3.3). The number of trials until the red-light flash elicited a positive response without any feeding reinforcement, tested after each ten trials, measured acquisition of the conditioned reflex. The number of deconditioning trials, consisting of red-light flash without feeding, until no positive response was observed measured extinction of the conditioned reflex.

Table 3.3. Results of experiments on hormonal influence on memory formation and retention in rainbow trout

	Treament	Number of trials (no. of days in small boxes) until:			
		Acquisition of response		Extinction of response by all fish	
PRE SMOLTS	TSH N=11	15	[3]	225+ No extinction	[95+]
	Saline N=11	60	[10]	49	[8]
	Uninjected N=11	60	[13]	45	[6]
POST SMOLTS	TSH N=11	61	[10]	47	[8]
	Saline	60	[10]	40	[8]
	Uninjected N=11	63	[11]	48	[7]

Results of this pilot study showed two significant differences between TSH- and saline-injected fish (Table 3.3). (1) The period until acquisition of the conditioned response was shorter for TSH- than saline-treated fish (15 trials vs. 60 trials), and (2) Extinction of the conditioned response occurred within 49 trials (8 days) in saline-injected fish and was not achieved after 225 trials (90 days) in TSH-injected fish.

There are, however, problems inherent with the use of the conditioned-learning model in testing hormonal effects upon learning and memory. The most crucial question is whether the observed effects of thyroid hormones on acquisition and extinction of a response are indeed alterations in learning and memory, or whether they result from a hormonal effect on such factors as attention, activity, metabolic rate, appetite, or sensitivity to stimuli. These factors are difficult to separate from direct effects on the learning or memory-acquisition process. However, the fact that in these experiments retention extended well beyond the time that thyroid hormone levels had returned to normal and extinction was not observed in 3 months of deconditioning suggests that a change in the memory process had, indeed, occurred with manipulation of TSH levels. Moreover, when we repeated this work with post-smolt trout (Scholz, personal observations from studies conducted at Eastern Washington University) no differences were observed between TSH-injected or control fish. This points to an organizational, as opposed to activational, influence of thyroid hormones. Apparently exposure to the hormone before a critical period in development "permanently" alters the memory processes. After the critical period thyroid hormones do not act in the same fashion.

Also potentially important with respect to olfactory imprinting is the observation that cortisol injections alter olfactory discrimination in salmon as determined by electroencephalographic studies (Oshima and Gorbman 1966 b). ACTH has also been implicated in memory retention in birds and mammals (DeWied 1964, 1966, 1969, 1973, Bohus et al. 1973, Champney et al. 1976, Dornbush and Nikolovski 1976, Flood et al. 1976, Gold and Van Buskirk 1976, Gold and McGough 1977, Martin 1978, reviewed by Levine 1968). Since most of the reported effects of ACTH on memory appear to be activational influences (reviewed by Scholz 1980), we have focused our attention on thyroid hormones.

Chapter 4

Fluctuations in Hormone Levels During the Spawning Migration: Effects on Olfactory Sensitivity to Imprinted Odors

In addition to the smolt stage, the salmon endocrine system also undergoes profound readjustments during the spawning migration. To understand our investigations on endocrine regulation of olfactory imprinting, which are presented in Chapter 5, some of these fluctuations in hormone levels during spawning must be considered.

Sex hormones increase dramatically during the course of the spawning season in salmonids. Radioimmunoassays of serum estradiol 17-β in female rainbow trout showed that the level of this hormone increased from basal levels of 125 pg ml^{-1} in July to 4,800 pg ml^{-1} in October during the peak of the spawning season (Whitehead et al. 1978; Fig. 4.1). In these experiments trout were held under different daylength regimes for several months. The data we have presented were for the simulated conditions that corresponded to the natural daylength. Under those conditions Whitehead et al. (1978) noticed a sharp increase in estradiol 17-β immediately after daylength began to decrease. In other experiments in which fish were subjected in increasing daylengths followed by decreasing daylengths, but out of phase with natural photoperiod, estradiol levels increased shortly after the shift to shorter daylengths. Whitehead et al. (1978) concluded that photoperiod is the major factor in the environmental control of reproductive activity in rainbow trout.

Spawning of salmon (coho, pink, Atlantic) and trout (rainbow, brown) has been advanced by decreasing daylength (Hoover and Hubbard 1937, Allison 1951, Hazard and Eddy 1951, Henderson 1963, Carlson and Hale 1973, de Vlaming 1974, MacKinnon and Donaldson 1976, reviewed by Poston 1978). In contrast, spawning was delayed in trout exposed to artificially prolonged daylength through the fall (Hoover and Hubbard 1937, Allison 1951, Hazard and Eddy 1951, Henderson 1963).

Photoperiodic cues stimulate the spawning migration of Pacific and Atlantic salmon by activating gonadotropin production. Plasma levels of gonadotropins, as determined by radioimmunoassay, increase in fish exposed to gradually decreasing daylengths (Breton and Billard 1977, Crim et al. 1976). Gonadotropin stimulates differentiation of gonads, induces production of sex steroid hormones, and induces migration. Injection of purified salmon gonadotropin or carp-pituitary extract is reported to in-

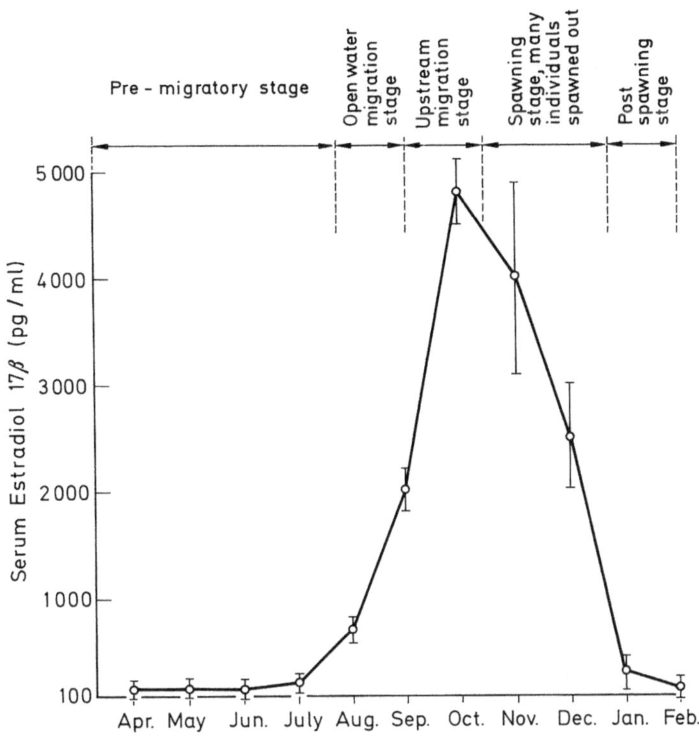

Fig. 4.1. Radioimmunoassays of estradiol 17-β in rainbow trout over the course of the spawning migration. (After Whitehead et al. 1978)

crease the weight of the gonads in juvenile and pre-spawning salmon, increase the incorporation of yolk proteins, or vitellogenin, into oocytes (i.e., oocyte maturation) in females, and increase spermatogenesis and shedding of motile spermatoza in juvenile or pre-spawning males (Robertson and Rinfret 1957, Schmidt et al. 1965, Donaldson et al. 1972, Funk and Donaldson 1972, Funk et al. 1973). In this connection, the most important investigations have been conducted by Edward M. Donaldson at the Canadian Government Fisheries and Marine Service Laboratory in Vancouver, British Columbia. After isolating gonadotropin (SG-G100) from the pituitary glands of spawning chinook salmon (Donaldson et al. 1972), he injected the preparation into juvenile pink and coho salmon and found that after 8–12 weeks of injection males became sexually mature (Donaldson et al. 1972, Funk and Donaldson 1972, MacKinnon and Donaldson 1978). Gonadotropin administration also increased the circulating levels of sex hormones as determined by radioimmunoassay (Crim and Evans 1976, Scholz 1980).

We have also been able to demonstrate that gonadotropin administration stimulated migratory behavior (Scholz 1980). Injection of Donaldson's SG-G100 gonadotropin into coho salmon at a non-migratory stage increased locomotor activity and upstream movement compared with saline-injected fish. As the injection also increased the level of sex hormones (from RIA determinations), it is not clear if the gonadotropin or the sex hormones are responsible for this altered behavioral state. In addition to their effects on behavior, SG-G100 injections induced the development of spawning coloration – red in males and purple in females – and the development of a spawning hook in juvenile male coho salmon. Presumably these secondary sexual characters were mediated by the increased level of sex hormones as injection of testosterone and estradiol 17-β are also reported to cause red color and bone deformation in the snout (Fagerlund and Donaldson 1969).

4.1 Effect of Sex Hormones on Olfactory Sensitivity and Discrimination

Sex hormones modify olfactory sensitivity and discrimination capability in salmon. Injections of estradiol or testosterone modify EEG activity in the olfactory bulb to odors, by altering threshold and, at a given concentration, amplitude (Oshima and Gorbman 1966a, 1968, Hara 1967a, b, Hara and Gorbman 1967, reviewed by Hara 1970). The effect observed was similar to that seen when certain regions of the brain are electrically stimulated. Hence, the hormones may bind to nerve cells and alter firing so as to exert centrifugal control over the olfactory system.

The effect of sex hormones on olfactory sensitivity is interesting in view of the fact that olfactory sensitivity of salmon to their imprinted odor increases over the course of the spawning season in concert with increased circulating levels of sex hormones. Olfactory sensitivity of coho salmon to their imprinted odor (morpholine) increased during the spawning migration, as indicated by increase in amplitude of the EEG response. The amplitude of the EEG from the olfactory bulb was recorded at intervals of one week from the beginning of the spawning season until immediately before spawning. Spawned-out fish did not react to the chemical (Cooper and Hasler 1973, Scholz et al. 1973; Fig. 4.2). The response of control fish to morpholine did not increase.

Oshima et al. (1973) found that the amplitude of the EEG response of chum salmon to their natural home water increased during the course of the spawning season. Salmon still in salt water had a low-amplitude EEG response to their home water, but by the time they were ready to begin the freshwater portion of their migration displayed a high-amplitude response

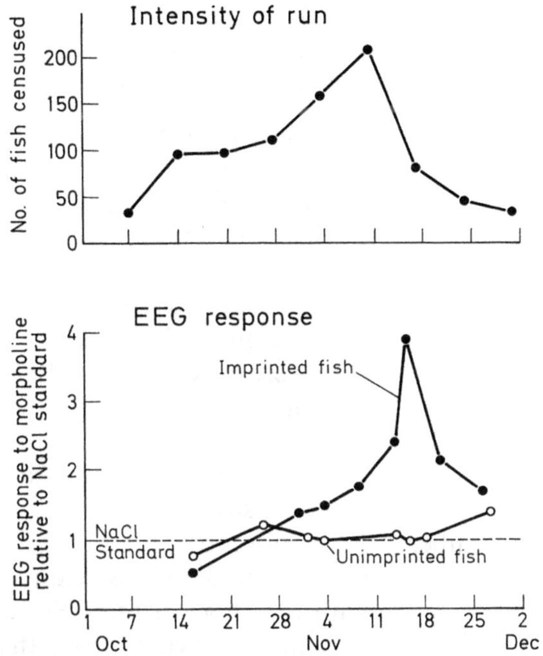

Fig. 4.2. EEG responses of morpholine-imprinted and control salmon to morpholine over the course of the spawning season. The amplitude of the response to morpholine by morpholine-imprinted fish increased over the course of the spawning season. Note that at the end of the spawning season morpholine imprinted fish stopped responding to morpholine. At this time they began to respond strongly to odors from other salmon. (After Cooper and Hasler 1973, Scholz et al. 1973)

to it. The fish had distinctive fin clips, so their home river could be ascertained even though they were captured in the ocean. Since fish were still in salt water when the amplitude increased, their response cannot be attributed to recent experience of the home stream. Water from non-home sources evoked a low amplitude response in both cases. Hence, before the upstream-migration phase, it appears that salmon cannot distinguish their home stream odor, but at about the time they reach the coast and begin migrating upstream, they can.

Unfortunately, in these two experiments, plasma levels of sex hormones were not measured, but in view of the influence of sex steroids on olfactory sensitivity as measured by EEG (reviewed by Hara 1970) and the documented increase in the level of sex hormones over the course of the spawning migration (Osborn et al. 1978), it seems reasonable to speculate that sex hormones might be involved in the "increasing sensitivity" of salmon to the imprinted odor.

In connection with this idea, we have made behavioral observations on the response of Lake Michigan coho salmon to their home-river odor over the course of the spawning season and have measured the levels of sex hormones in each individual by radioimmunoassay. This investigation was conducted in collaboration with Robert W. Goy, Director at the University of Wisconsin Regional Primate Center, in the fall of 1978.

We captured coho salmon from the Ahnapee River at five different times. The dates, our reasons for selecting these dates, locations of capture, and general comments about the status of the fish are as follows:

June: Fish were collected in Lake Michigan before the migration season to provide information about the basal level of sex hormones. Fish from the Ahnapee River could be identified by distinctive fin clips. As the fish were not in a migratory state, their behavior was not tested. The fish were silver-colored and gonad weight averaged less than 1% of the total body weight of the fish.

August: Fish were collected from Lake Michigan during the open water portion of their migration to correspond with those collected in the ocean for the EEG studies described above. Gonad weight was approximately 40% of body weight. The fish were silver-colored and not in reproductive condition.

September: Fish were captured at the mouth of their home river at the beginning of their upstream migration period to correspond with the stage in the EEG experiments where salmon were making the switch from the open-water phase to the upstream phase of their migration. Fish were in spawning coloration, but reproductive condition was variable. Eggs or milt could not be squeezed from most fish (i.e., hard). However, eggs or milt flowed freely from some when they were handled (i.e., ripe). Examination of ovaries showed that eggs were still contained within the ovarian sac in "hard" females. As the gonads mature the sac disintegrates and the eggs are free in the body cavity in "ripe" females.

October: Fish were captured in the home river at the peak of upstream migration. At this stage most fish were ripe but had not spawned.

November: Fish were captured in their home river at the end of the spawning season. Some fish were ripe but most fish were partially or completely spawned out.

After capture, the fish were transported to a small tributary located 60 km from their home river (Fig. 4.3) where behavioral tests were conducted. Our aim was to determine the responses of fish at various stages of migration, or with different plasma levels of sex hormones, to different concentrations of home-stream water introduced into the test tributary. Water from the home river was pumped with a portable water pump into a 4,000-l tank mounted on a truck and transported to the test tributary along with the fish. Tests were conducted by pumping water from the tank into the test tributary. Flow rate of the tributary was calculated by determining the

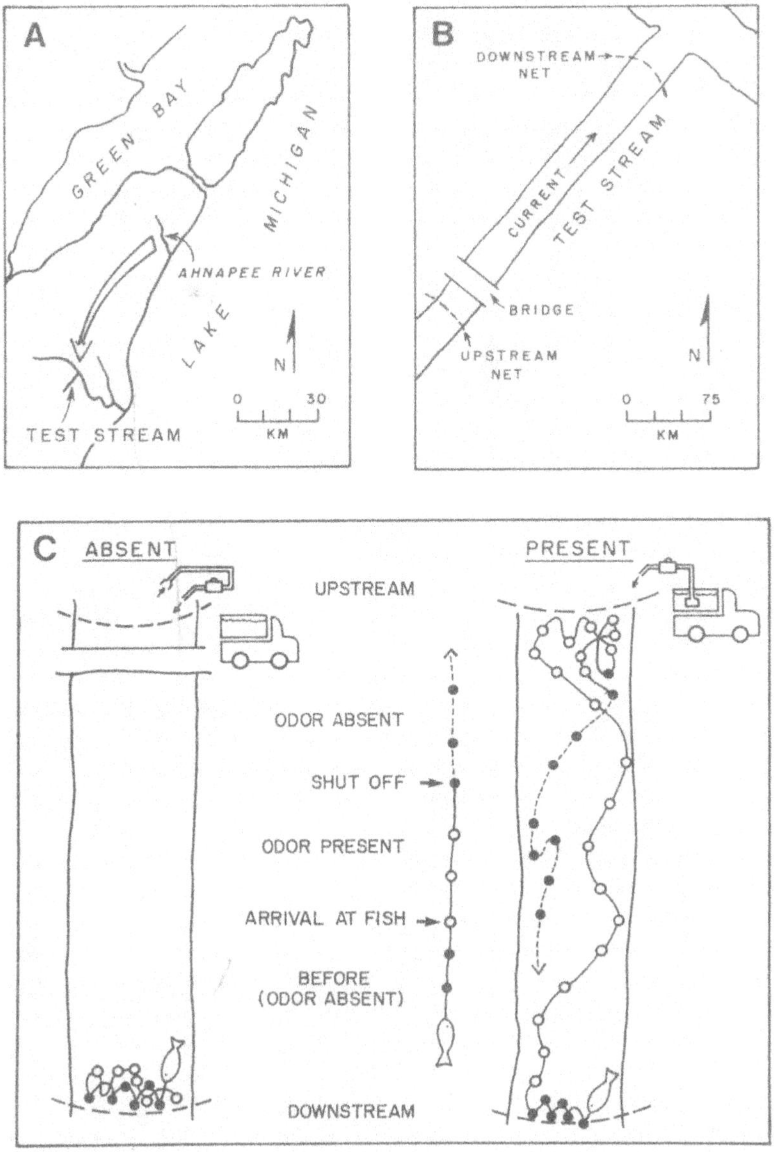

Fig. 4.3 A–C. Study area for tests of olfactory discrimination by coho salmon of home-stream water. Included are maps of (*A*) location of home stream in relation to test stream, (*B*) detail of study area, and (*C*) plots of one fish showing responses during control trials (*left*) and with home-river water present (*right*). Note the setting of the pump in the *upper right corner* in each case. *Dots* represent fish position at 15-s intervals

Table 4.1. Behavior and plasma levels of estradiol 17-β of female coho salmon collected at different intervals over the course of the spawning migration

Group	Fish no.	Color	Reproductive condition	E_2 level (ng ml^{-1} serum)	Behavior: Control (0%) Swimming speed (cm s^{-1})	Control (0%) Direction	Test (10%) Swimming speed (cm s^{-1})	Test (10%) Direction	Test (25%) Swimming speed (cm s^{-1})	Test (25%) Direction
June – Collected in lake. Gonad wt. 1% of fish body wt. (except for specimen no. 6=10% of body wt.)	1	Silver	–	0.08	Not in migratory state					
	2	Silver	–	0.10						
	3	Silver	–	0.11						
	4	Silver	–	0.10						
	5	Silver	–	0.05						
	6	Silver	+	1.00						
	7	Silver	–	0.05						
Aug. – Collected in lake during open water. Portion of migration Gonad wt. about 40% of fish body wt.	1	Silver	H	3.36	–	S	–	S	–	S
	2	Silver	H	0.57	7.8	D	8.4	D	–	S
	3	Purple	HR	16.43	–	S	–	S	17.5	U
	4	Silver	H	1.37	14.1	M	13.1	D	13.7	M
	5	Silver	H	1.85	12.1	D	12.4	M	11.3	M
	6	Silver	H	2.11	–	S	–	S	–	S
	7	Silver	H	4.12	–	S	–	S	–	S
Sept. – Collected in river of beginning of upstream migration. Gonad condition: eggs hard in sac, cannot force eggs out by squeezing	1	Silver	H	4.65	7.7	D	7.8	D	7.3	D
	2	Silver	H	5.23	6.5	D	6.8	D	23.4	U+
	3	Purple	R	10.27	13.1	M	17.8	M	24.8	U+
	4	Silver	HR	11.38	–	S	–	S	–	S
	5	Silver	H	5.49	7.8	D	9.2	D	29.8	U

Oct. – Collected in river at peak of upstream migration. Many fish release milt when picked up or squeezed	1	Red	R	55.4	13.4	D	27.8	U		
	2	Red	R	23.08	7.1	D	24.1	U+		
	3	Red	R	47.31	–	S	–	S	–	S
	4	Red	R	38.88	14.7	D	27.8	U+		
	5	Red	R	52.99	16.7	D	28.9	U+		
	6	Red	R	37.58	11.8	D	26.5	U		
	7	Red	R	36.29	7.4	D	26.4	U+		
Nov. – Collected in river at end of spawning season. Many fish spawned out or partially spawned out	1	Coal	P.S.O.	18.71	–	S	–	S	–	S
	2	Red	R	39.04	11.7	D	27.1	U+		
	3	Red	R	56.80	7.9	M	8.7	D	8.3	D
	4	Red	R	76.84	10.3	D	24.7	U		
	5	Red	R	50.27	8.4	D	26.3	U+		
	6	Coal	S.O.	10.41	–	S	–	S	–	S
	7	Coal	S.O.	0.47	–	S	–	S	–	S
	8	Red	R	42.31	13.1	D	28.8	U+		
	9	Coal	S.O.	8.21	–	S	–	S	–	S

H = hard; R = rip; P.S.O. = partially spawned out; S.O. = spawned out; S = stationary; D = downstream; U = upstream; U+ = swam upstream and localized source of odor

Table 4.2. Behavior and testosterone levels of male coho salmon collected at different intervals over the course of the spawning migration

Group	Fish no.	Color	Repro-ductive con-dition	T level (ng ml^{-1} serum)	Behavior					
					Control (0%)		Test (10%)		Test (25%)	
					Swimming speed (cm s^{-1})	Direction	Swimming speed (cm s^{-1})	Direction	Swimming speed (cm s^{-1})	Direction
June – Collected in lake.	1	Silver	–	0.77	Not in migratory state					
Gonad wt. 1% of fish	2	Silver	–	0.92						
body wt. No spawning	3	Silver	–	0.47						
hook present.	4	Silver	–	1.06						
Aug. – Collected in lake.	1	Red	H	14.41	–	S	–	S	–	S
During open water	2	Silver	H	12.21	–	S	–	S	–	S
portion of migration.	3	Silver	H	7.65	7.6	D	14.1	M	8.9	M
Gonad wt. about 40%	4	Silver	H	10.58	–	S	–	S	7.8	M
of fish body wt. Spawning	5	Silver	H	11.43	13.1	M	14.7	M	14.9	M
hooks beg. to develop	6	Silver	H	11.73	–	S	–	S	–	S
Sept. – Collected in riv. at beg.	1	Red	H	13.57	–	S	–	S	–	S
of upstream migration; Gonads	2	Red	HR	17.56	–	S	–	S	14.8	U
firm; do not release milt if	3	Red	R	18.91	13.2	D	10.8	D	27.4	U+
squeezed; spawning hooks well dev.	4	Red	HR	17.51	–	S	–	S	–	S

Oct. – Collected in river at peak of upstream migration. Gonad condition: some fish w/eggs in sac; others w/eggs in body cavity. Eggs can be forced out by squeezing	1	Black	R	23.72	7.8	D	27.1	U		
	2	Black	R	24.06	14.1	D	34.8	U+		
	3	Purple	R	19.78	25.8	D	28.2	U+		
	4	Purple	R	12.48	13.1	M	19.5	U+		
	5	Black	R	14.81	11.4	D	28.2	U		
	6	Purple	R	14.56	9.2	M	25.6	U+		
Nov. – Collected in river at end of upstream migration. Many fish spawned out or partially spawned out	1	Coal	S.O.	1.67	–	S	–	S	–	S
	2	Black	S.O.	2.29	–	S	–	S	–	S
	3	Black	R	23.67	9.1	D	24.7	U+		
	4	Black	R	24.79	13.7	D	25.1	U		
	5	Black	R	24.36	7.7	M	28.7	U+		
	6	Coal	P.S.O.	10.62	–	S	–	S	–	S

H = hard; R = ripe; P.S.O. = partially spawned out; S.O. = spawned out; S = stationary; D = downstream; U = upstream; U+ = swam upstream and localized source of odor cross-sectional area of the channel and measuring water-current velocity with an electronic current meter. The introduction of Ahnapee River water into the test stream was adjusted to produce concentrations of either 10% or 25% for intervals of 5 min.

We blocked off a 200 m section of the tributary by stretching nets across it and released the fish individually into this enclosure. A fishing float was attached to the base of the dorsal fin to permit visual tracking.

Ahnapee River water was pumped in at the upstream net. Based on the experiments of Peter Johnsen (described in Chap. 2), we expected that if the home-river odor were absent salmon would migrate downstream, i.e., the absence of the home-stream odor would evoke negative rheotaxis. If a pulse of home-river water were entered, we expected the fish to orient to it and migrate upstream since presence of the odor would act as a sign stimulus for releasing positive rheotaxis. Before monitoring the effect of home-river water, we conducted sham trials by pumping tributary water into the creek. This controlled for the possibility that salmon move upstream as a result of some facet of the experimental procedure not related to odor, e.g., increased water-current velocity. Thus, a display of upstream migratory behavior in the presence of home-river water would indicate that the fish were sensitive to it. By comparing the relative responsiveness of fish to two different concentrations of home-river water over the course of the spawning season, we could determine if sensitivity increased as the spawning season progressed. By measuring sex-hormone levels of individual fish, we could ascertain whether olfactory sensitivity is correlated with increasing levels of sex hormones.

After a 4-h acclimation period, we monitored the behavior of the fish during three 5-min control (pretest) periods by recording their positions at 15-s intervals on a map of the enclosure. This information was quantified and translated into average swimming speed and overall course direction (upstream, stationary, downstream). Then tests with 10% and 25% home-river odor, respectively, were conducted, and the responses quantified. Results are presented in Tables 4.1 and 4.2.

Immediately after the behavioral tests, blood samples were collected by heart puncture and the fish were killed for determination of sex and gonad condition. Blood samples were analyzed for testosterone in males and estradiol 17-β in females by radioimmunoassay.

Results demonstrated that sensitivity or discrimination of home-river water is correlated with sex-hormone level in both male and female subjects. After the acclimation period all the fish were either actively trying to swim downstream, i.e., they swam back and forth along the downstream net, or were stationary at the downstream end of the enclosure. Pumping of (control) water did not alter this activity. Addition of either 10% or 25% Ahnapee River water induced upstream migration in some cases:

1. In August 0 to 13 fish responded to 10% and 1 of 13 to 25% Ahnapee River water: average testosterone level in males was 11.3 ng ml^{-1}; average estradiol level in females was 4.25 ng ml^{-1}.
2. In September 0 of 9 and 5 of 9 fish reacted respectively to 10% and 25% home river water: average testosterone levels in males was 16.89 ng ml^{-1}; average estradiol level in females was 7.40 ng ml^{-1}.
3. In October 12 of 13 fish responded to 10% Ahnapee River water: average testosterone level in males was 41.64 ng ml^{-1}, average estradiol level in females was 17.40 ng ml^{-1}.
4. In November partially spawned out or spawned out fish with low steroid levels (average testosterone level in males = 6.36 ng ml^{-1}; average estradiol level in females = 4.68 ng ml^{-1}) did not react to either concentration of river water (0 of 4 males and 0 of 3 females) while those which were still ripe with high steroid levels (testosterone in males = 47.32 ng ml^{-1}; estradiol in females = 24.20 ng ml^{-1}) did (5 of 5 males and 3 of 3 females).

In summary, sensitivity of the fish to Ahnapee River water apparently increased over the course of the spawning season. During this period there was a sharp increase in the level of sex hormones. Thus, behavior did correlate with difference in sex hormone levels. Also, one female fish (no. AUG-3) collected at the beginning of the spawning season but with above average levels (16.43 compared with the average of 4.25 ng ml^{-1}) responded to Ahnapee River water like females tested in October (average estradiol = 17.40 ng ml^{-1}). In addition, at the end of the spawning season, spawned-out fish with low levels of sex hormones did not respond to their home-stream odor, while those that had not yet spawned and still had high levels of sex hormones did.

It is also possible that fish might respond more strongly to home-river water simply because the characteristics of the water changed over the course of the spawning season. For example, as increasing numbers of spawning adults were entering the river, more of their odor would be added to the water. Therefore, water from the home stream was collected upstream from a dam, i.e., from an area that was inaccessible to returning salmon, so that in our study it is unlikely that the odor characteristics of the water changed over the course of the spawning season due to factors such as conspecific odors produced by returning fish. Also, the fact that fish tested with the same water sample at the same time responded in different fashion to home-river water, and that those reacting to the odor all had high sex-hormone levels, while those not reacting had low levels, strongly suggests that response to the odor was related to hormone level and not to seasonal changes in characteristics of the water.

4.2 Physiological Mechanism for Countering Olfactory Adaptation

Information about increasing olfactory sensitivity in spawning salmon may be important from the standpoint of sensory adaptation to the home-stream odor. Johnsen (1978) pointed out that in his experiments salmon followed the edge of an odor trail and suggested that it may reflect a behavioral mechanism for preventing adaptation. While this might apply in situations near the natal tributary where the water from the tributary might be flowing as a distinct water mass, it seems unlikely that such a behavioral mechanism could operate farther downstream where home-tributary water would be mixed horizontally and vertically. The fact that many salmon migrate hundreds or thousands of kilometers upstream serves to increase our skepticism about the general applicability of a behavioral mechanism for prevention of adaption. About the only behavioral system that we could envision would be for the fish to clear its nasal chambers by bringing them out of the water. Since salmon often break water during their upstream migration, this is a distinct possibility; however, we favor a physiological mechanism for prevention of adaptation. If olfactory sensitivity to the imprinting odor continues to increase during the spawning migration, it may be that olfactory adaptation would not be a problem. The underlying physiological mechanism could be the increased levels of sex hormones.

In mammals sex hormones fluctuate from basal levels of about 150 pg ml^{-1} to peaks of 350–1,000 pg ml^{-1} during reproductive cycles (data for estradiol 17-β in Rhesus monkeys). In coho salmon, in a 1-month period when the fish are migrating upstream, estradiol 17-β increases from about 500 pg ml^{-1} to 18,000 pg ml^{-1} or about 600 pg day ^{1}, i.e., about the same fluctuation as an entire reproductive cycle in mammals in 1 day! Thus, it seems likely that increased levels of sex hormones could exert profound physiological effects on the olfactory system.

These studies affected our thinking in planning experiments designed to elucidate the effect of manipulation of hormone levels upon the imprinting process because they demonstrate that fish not in spawning condition might not be expected to react to a chemical to which they had previously been imprinted.

Another important point about the work described in this chapter is that salmon stop reacting to their imprinting odor at about the time of spawning and begin responding strongly to the odor of other salmon. We view this as case of adaptive behavior which could act as a "safety-valve" mechanism to insure spawning even if the fish did not home correctly. If a fish has failed to home or becomes lost because of problems with its imprinting mechanism, it could be attracted into a tributary where other spawning salmon are present.

Chapter 5

Thyroid Activation of Olfactory Imprinting in Coho Salmon

From the work reviewed here (Chap. 3), we formulated the hypothesis that olfactory imprinting is facilitated by increases in levels of thyroid hormones at the time of smolt transformation (Scholz 1980). Our overall plan for testing this hypothesis was to inject TSH or ACTH into pre-smolt coho salmon while simultaneously exposing them to a synthetic chemical, either morpholine or phenethyl alcohol. Ten months later behavioral tests were performed to determine if the fish had retained a long-term olfactory memory of the odor to which they had been exposed at the time of hormone treatment. Although this experiment is simple conceptually, it proved to be difficult to perform because of the controls that were required. In performing this set of experiments we had three major objectives:

Objective 1. To define operationally smolt transformation by quantifying morphological, physiological and behavioral transitions at 4-week intervals from January through June, i.e., pre-smolt through smolt stages, and correlating these transitions with seasonal fluctuations in circulating levels of thyroid hormones and cortisol. The indicators of smolt transformation that we measured included levels of triiodothyronine, thyroxine and cortisol, degree of silvery coloration, downstream migratory activity, salinity tolerance, osmoregulatory capability and gill $Na+/K+$ ATPase activity. These provided baselines for comparison with experimental fish.

Objective 2. To assess the effect of hormone injections on inducing smolt transformation. This was an important control procedure because if the hormone injections did not induce smolt transformation then they would not be expected to have an effect on olfactory imprinting. Experimental protocol involved injecting pre-smolts with: (a) TSH until thyroxine and triiodthyronine levels approximated those observed in natural smolts; (b) ACTH until cortisol levels approximated those of natural smolts; (c) both TSH and ACTH; (d) saline and (e) uninjected. Saline and uninjected fish served as controls for general effects of handling and injection. Hence, since we worked with physiological as opposed to pharmacological concentrations of the hormones, we felt that we were simply accelerating smolt transformation. Before and after the hormone treatment we assessed the morphological, physiological, and behavioral transitions by quantifying them in the same manner used for natural smolts. Hence, we could determine if transitions in experimental fish actually resembled those that occur during natural smolt transformation, and also evaluate the effects of specific hormones upon each of the morphological, physiological, and behavioral transitions.

Objective 3. To determine whether TSH or ACTH injections, along with simultaneous exposure to either morpholine or phenethyl alcohol, resulted in imprinting to that odor. Saline-injected and uninjected fish, also exposed to the synthetic chemicals, controlled for the possibility that the fish were generally sensitive to the chemicals independently of the hormone injection. The hormone treatment continued until the fish appeared to undergo smolt transformation, after which it was stopped and the supply of synthetic chemicals was withdrawn. The fish were then held for 10 months before conducting behavioral and physiological tests to determine if they could discriminate their treatment odor. Odor discrimination tests were conducted in a natural river system below the confluence of two tributaries. After morpholine or phenethyl alcohol was introduced into either arm, the fish were released 150 m below the junction. A positive response required upstream migration and selection of a tributary scented with the correct treatment odor. Hence, the problem posed to the fish was similar to the problem faced by a naturally migrating adult salmon. Before we conducted the tests, the fish were injected with gonadotropin to bring them into a migratory disposition and to mimic the physiological state of naturally spawning salmon. This was done because of the information presented in Chapter 4 about the effect of sex hormones on olfactory discrimination of imprinted odors by salmon. As an additional control measure, we also exposed natural smolts held at a hatchery to morpholine or phenethyl alcohol as in our artificial imprinting experiments described in Chapter 2, and tested these fish along with the experimental subjects in the olfactory discrimination tests. From our earlier work, we assumed that these hatchery controls would become imprinted; therefore, their behavior toward the imprinting chemical provided a baseline for comparison with experimental fish.

5.1 Methods for Studying Hormonal Regulation of Smolt Transformation

The following methods were used for studying hormonal regulation of smolt transformation in natural and experimental smolts.

5.1.1 Natural Smolts

Plasma levels of thyroid hormones and cortisol from coho salmon undergoing natural smoltification were monitored by radioimmunoassay (Scholz 1980). Blood samples were collected at approximately 4-week intervals from the pre-smolt stage in January until smolt transformation occurred in May, and at each interval, morphological, physiological and behavioral transitions were quantified. The fish were obtained from a fish hatchery where they were held in outdoor raceways, i.e., exposed to natural photoperiod and constant-temperature spring water of 6 °C until used for experiments. Hormone assays and the test procedures for quantifying degree of silvering, downstream migratory activity, salinity tolerance, osmoregulatory capability and Na^+/K^+-ATPase activity in gills are described in detail below.

5.1.2 Hormonally Induced Smolt Transformation

Experimental fish were brought to the laboratory in January and maintained under constant photoperiod (8L : 16D) and constant temperature (6 °C) in two artificial raceways (i.e., 600 l Living Streams Systems; Frigid Units, Inc. Models LS 700 with D1-33 water-chiller pump unit). A different synthetic chemical was introduced into each tank. In each tank, pre-smolt fish were subdivided into five lots, each marked with a distinctive fin clip that corresponded to its treatment.

Fish were given injections of TSH, ACTH, TSH and ACTH, saline (controls) or left uninjected (controls) for 3 week in January, until thyroid hormone or cortisol levels in the hormone-treated subjects approached the levels observed for natural smolts. Before and after hormone treatment the degree of smoltification was assessed by measuring amount of silvering, downstream migratory activity, salinity tolerance, osmoregulatory capability, and Na^+/K^+-ATPase activity. The tests used were conducted in the same manner as those for natural smolts. Induction of morphological, physiological, and behavioral transitions that normally occur during smolt transformation would indicate hormonal involvement in regulating these processes provided that they did not also occur in saline or uninjected control fish. The degree to which transitions in experimental fish actually resembled processes occurring in natural smoltification was determined by quantitative comparisons between experimental and natural smolts.

5.1.3 Quantification of Smolt Transitions

Morphological, physiological, and behavioral transitions were quantified in natural smolts and experimental fish by using the same techniques. For additional details of these tests see Scholz (1980).

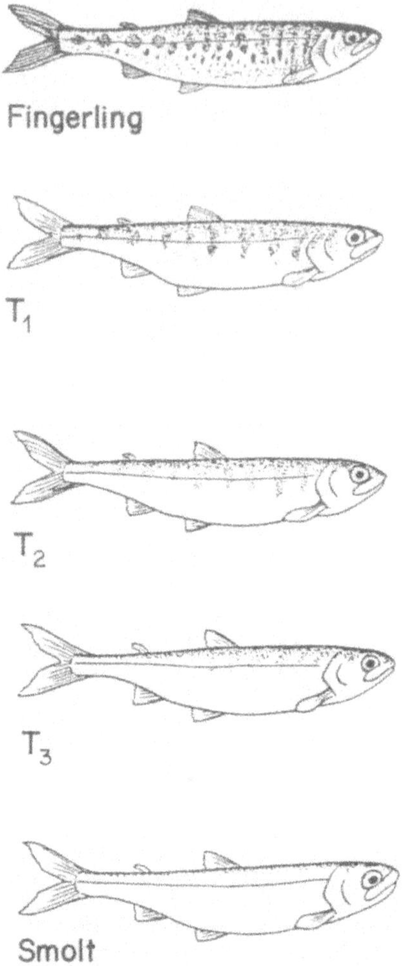

Fig. 5.1. Scheme used for categorizing silvering and loss or parr marks. T_1–T_3 = transition phases. Note how the parr marks disappear first from the lateral line, then from the anterior and posterior ends

5.1.4 Silvering and Loss of Parr Marks

Body coloration was determined by making visual observations and taking photographs of the fish. The fish were classified into categories (Fig. 5.1) described by Chrisp and Bjorn (1978):

Parr – dark brownish or yellowish green with distinct parr marks.
Transition phases – turning silver (T_1) parr marks beginning to disappear below the lateral line (T_2) and then from the head or tail (T_3).
Smolts – silver, parr marks completely gone.

The number of fish in each category at each time interval for natural smolts, or before and after injection in experimental fish, was recorded.

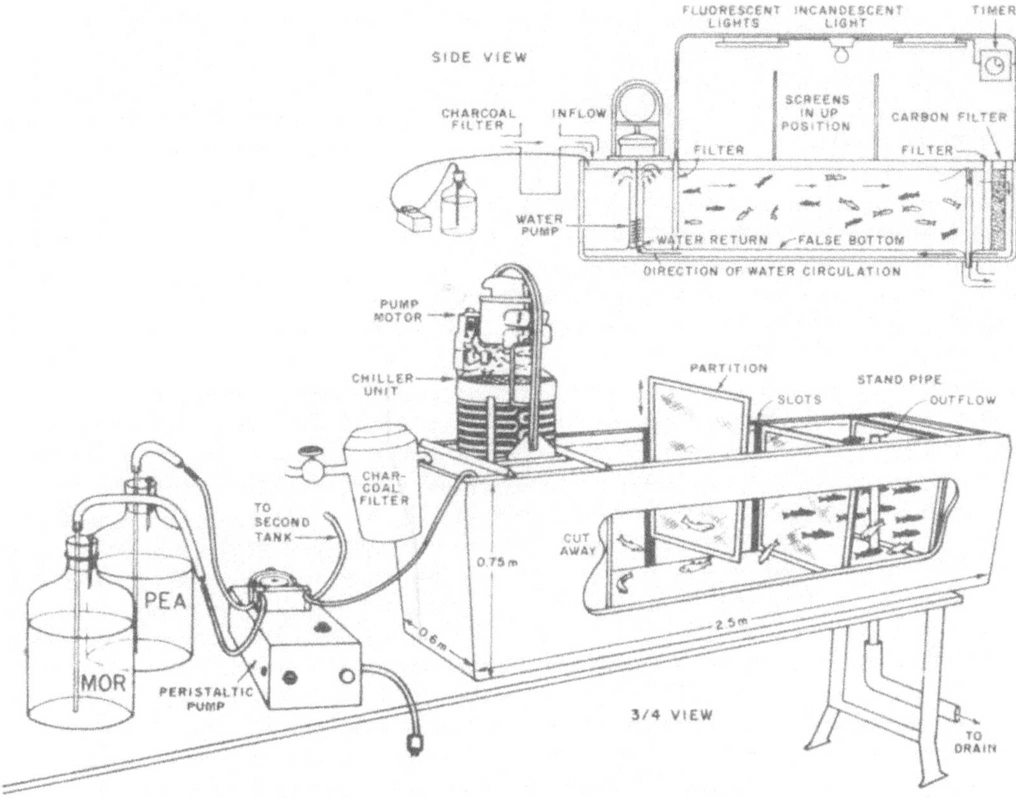

Fig. 5.2. Living Stream and odor delivery system. Note how the flow-through circulation creates upstream and downstream ends. Two screens could be fitted into slots and pushed down, trapping fish in the upstream, middle, or downstream compartments. The number of fish from each treatment group in each compartment were counted to assess downstream migratory activity. TSH- and saline-treated fish are represented respectively in *black* and *white*. *Side view at the top* of the diagram indicates that all fish were randomly distributed at the beginning of the experiment. In the $^3/_4$ *view on the bottom*, TSH-treated fish are orienting downstream, whereas saline-treated fish are randomly distributed after 3 weeks of injections

5.1.5 Downstream Migratory Behavior

Downstream migratory activity in natural and experimental fish was tested by placing them in Living Stream systems [1 m deep × 1 m wide × 2.5 m long refrigerated tanks with flow-through circulation (Fig. 5.2)]. In this system a water pump at one end lifts water from under a false bottom and recirculates it. A stand pipe drain is located at the opposite and so the net effect is that water flows through the tank in one direction. Charcoal-filtered tap water (Madison City supply) was introduced continuously at the upstream end at a rate of 2.5 l min^{-1}.

The Living Streams were equipped with screens that could be used to partition them into three compartments. To determine downstream migratory activity, the screens were fitted into their slots and left raised for 1 h, after which they were pushed down, trapping the fish in the upstream, middle, or downstream sections. Each type of fish in each compartment was then counted. Because salmon normally migrate at night, tests were conducted in the dark between 22.00 and 03.00 h. The test was repeated ten times for each group of fish tested. The mean and standard deviation(s) for the number of fish from each experimental group occupying the downstream compartment were recorded.

5.1.6 Salinity Tolerance and Osmoregulatory Capability

Salinity tolerance was determined by placing fish in 10‰, 20‰, or 30‰ seawater (Instant Ocean Salts) and calculating percent survival after 96 h. Osmoregulatory capability was determined by placing fish in freshwater, 10‰, 20‰, or 30‰ salt water for 2–96 h, after which osmotic concentration of blood serum and environmental water were measured with a vapor-pressure osmometer.

5.1.7 Gill Na^+/K^+-ATPase Activity

Na^+/K^+-ATPase specific activity was measured by the whole homogenate method described by Ewing and Johnson (1976), Johnson et al. (1977), and Scholz (1980).

5.1.8 Hormone Radioimmunoassay

Blood (0.2–0.5 ml) was collected by heart puncture, centrifuged for 10 min at 1,000 rpm and serum extracted and stored at −20 °C in glass vials until assayed. Because of the possibility of diurnal fluctuations in hormone

Fig. 5.3 A–F. Transitions in hormone levels, physiology and behavior during natural smoltification. *A* Hormone concentrations as determined by radioimmunoassay; n = number of individuals tested (each assayed in duplicate). *Brackets* indicate mean and standard deviation; (*T-3*) triiodothyronine; (*T-4*) thyroxine, (*C*) cortisol. *B* Body coloration. Each *point* represents one individual. *C* Downstream migratory behavior: percent in downstream compartment. Each *point* represents the mean and standard deviation of ten trials conducted at that time. *D* Salinity tolerance. Percent survival after 96 h in 10‰, 20‰, and 30‰. *E* Na^+/K^+-ATPase activity in gills. *F* Osmoregulatory capability. Osmotic concentration of blood serum was determined for fish held in freshwater, 10‰, 20‰, and 30‰ and plotted against osmotic concentration (salinity) of the medium it was held in. n = number of fish tested at each salinity. *Diagonal line* is the isosmotic line

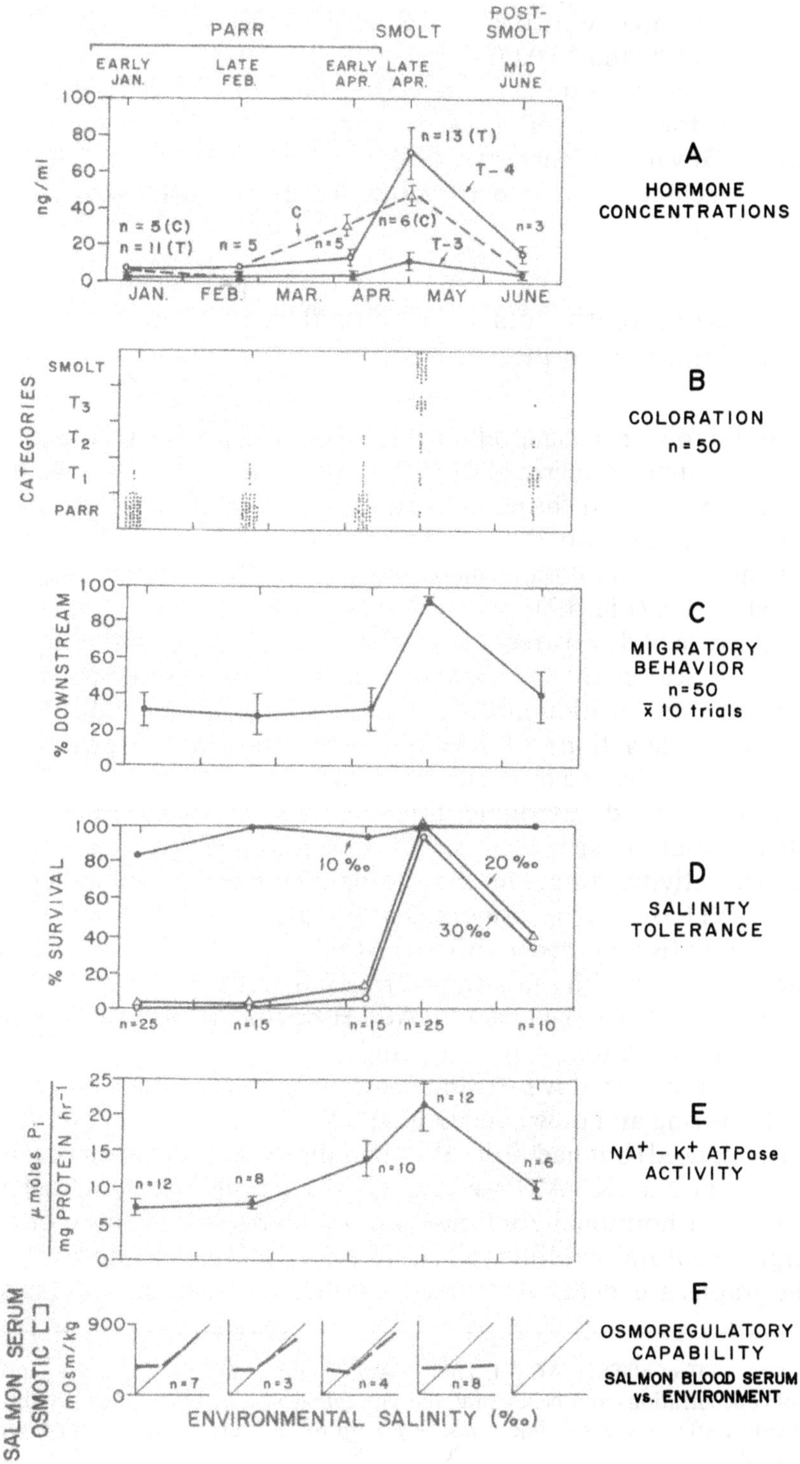
PARR
SMOLT
POST-SMOLT
EARLY JAN.
LATE FEB.
EARLY APR.
LATE APR.
MID JUNE
ng/ml
n = 13 (T)
T-4
C
n = 6 (C)
n = 5 (C)
n = 11 (T)
n = 5
n = 3
T-3
JAN. FEB. MAR. APR. MAY JUNE
A
HORMONE CONCENTRATIONS
CATEGORIES
SMOLT
T3
T2
T1
PARR
B
COLORATION
n = 50
% DOWNSTREAM
C
MIGRATORY BEHAVIOR
n = 50
x̄ 10 trials
% SURVIVAL
10 ‰
20 ‰
30‰
n = 25 n = 15 n = 15 n = 25 n = 10
D
SALINITY TOLERANCE
µmoles Pi / mg PROTEIN hr-1
n = 12
n = 8
n = 10
n = 6
E
NA+ - K+ ATPase ACTIVITY
SALMON SERUM OSMOTIC [ ]
mOsm/kg
900
n = 7
n = 3
n = 4
n = 8
ENVIRONMENTAL SALINITY (‰)
F
OSMOREGULATORY CAPABILITY
SALMON BLOOD SERUM vs. ENVIRONMENT

levels, all the samples were collected at approximately the same time of day, i.e., between 09.00 and 10.00 h.

Hormone levels were measured by standard radioimmunoassays (see Scholz 1980 for details of these particular assays) in the laboratory of Robert W. Goy at the University of Wisconsin Primate Center. Hormones monitored included thyroxine, triiodothyronine, cortisol, testosterone, and estradiol-17-β.

5.2 Results of Experiments on Hormonal Regulation of Smolt Transformation

In natural smolts, plasma levels of thyroxine, triiodothyronine and cortisol levels were found to increase from 2 to 10, 10 to 75 and 1 to 40 ng ml^{-1} serum, respectively, at the end of April just before development of silvery coloration, increased downstream migratory activity, salinity tolerance, osmoregulatory capability and increased gill Na^+/K^+-ATPase activity were observed in May (Fig. 5.3).

In the case of downstream migratory activity, pre-smolts were generally randomly scattered in all three compartments and were highly variable in terms of their position in the tank at any one time (as indicated by the high standard deviation), while smolts were always at the downstream end of the tank (as indicated by the low standard deviation).

Levels of triiodothyronine, thyroxine and cortisol declined in post-smolts to about pre-smolt levels. This was accompanied by a reduction in migratory activity, reversion back to parr coloration and reductions in Na^+/K^+-ATPase activity, osmoregulatory capability and salinity tolerance.

Pre-smolt fish injected with TSH, ACTH, or both displayed various attributes associated with smolt transformation, but saline or uninjected control fish did not (Figs. 5.4 and 5.5). ACTH treatment increased salinity tolerance, osmoregulatory capability, and Na^+/K^+-ATPase activity in gills, but not silvering or downstream migratory activity. TSH treatment increased silvering and downstream migratory behavior (standard deviation was also reduced) but had little effect on salinity tolerance, osmoregulatory capability, or Na^+/K^+-ATPase activity. The physiological and behavioral transitions in hormonally injected pre-smolts resembled those of salmon undergoing natural smoltification. (Note that in Figs. 5.3 and 5.4 the scales on the graphs are identical so that quantified measurements of physiolog-

Fig. 5.4 A–F. Effect of TSH, ACTH, TSH + ACTH and saline injections on T_3 and T_4 and cortisol concentrations and behavioral and physiological transitions. *B* before the start of treatment. *A* after 3 weeks of injections. *A–F* as in Fig. 5.3, except in *B* each *dot* represents two individuals

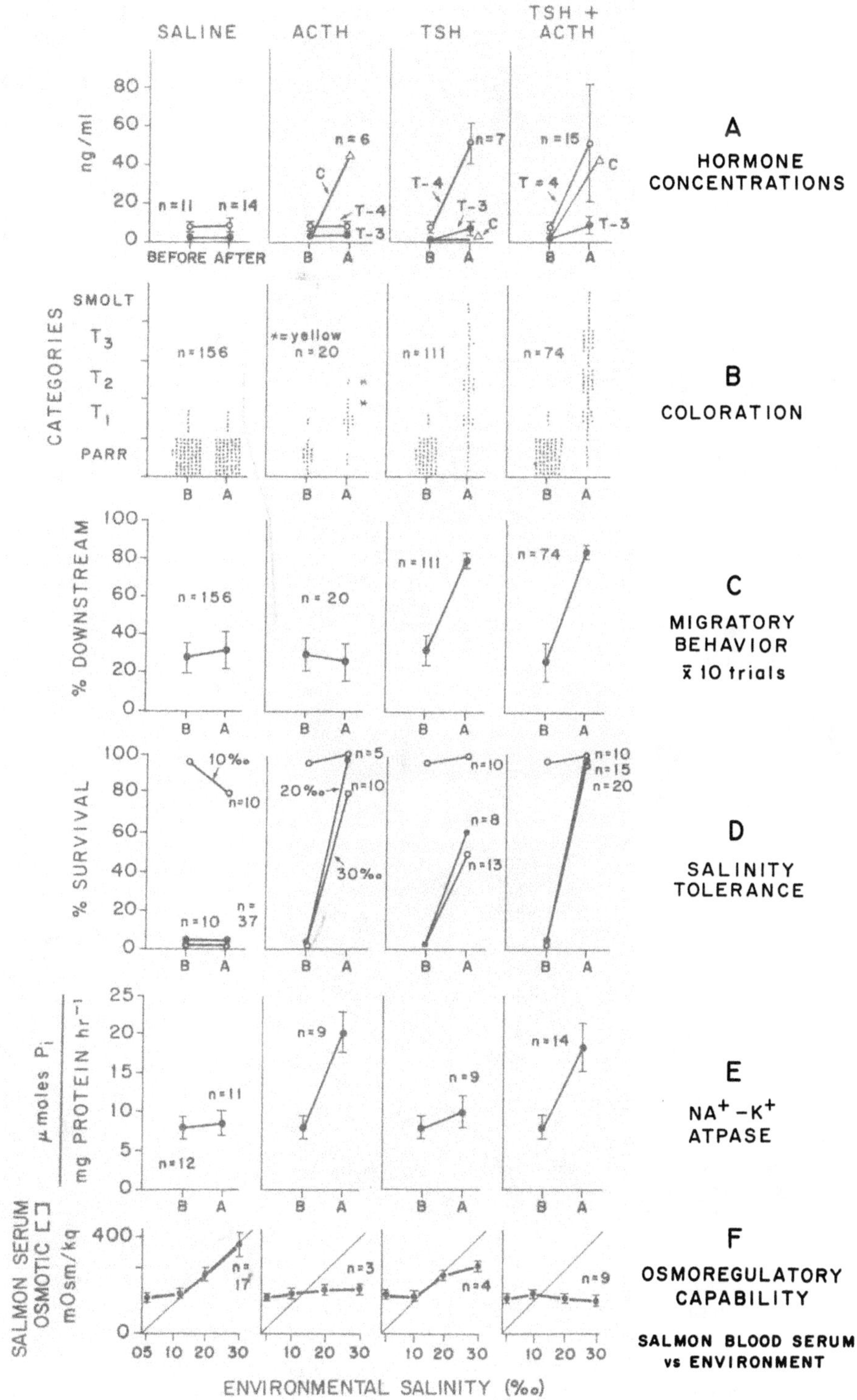

SALINE
ACTH
TSH
TSH + ACTH
A
HORMONE CONCENTRATIONS
B
COLORATION
C
MIGRATORY BEHAVIOR
x̄ 10 trials
D
SALINITY TOLERANCE
E
NA+ -K+ ATPASE
F
OSMOREGULATORY CAPABILITY
SALMON BLOOD SERUM vs ENVIRONMENT
ENVIRONMENTAL SALINITY (‰)

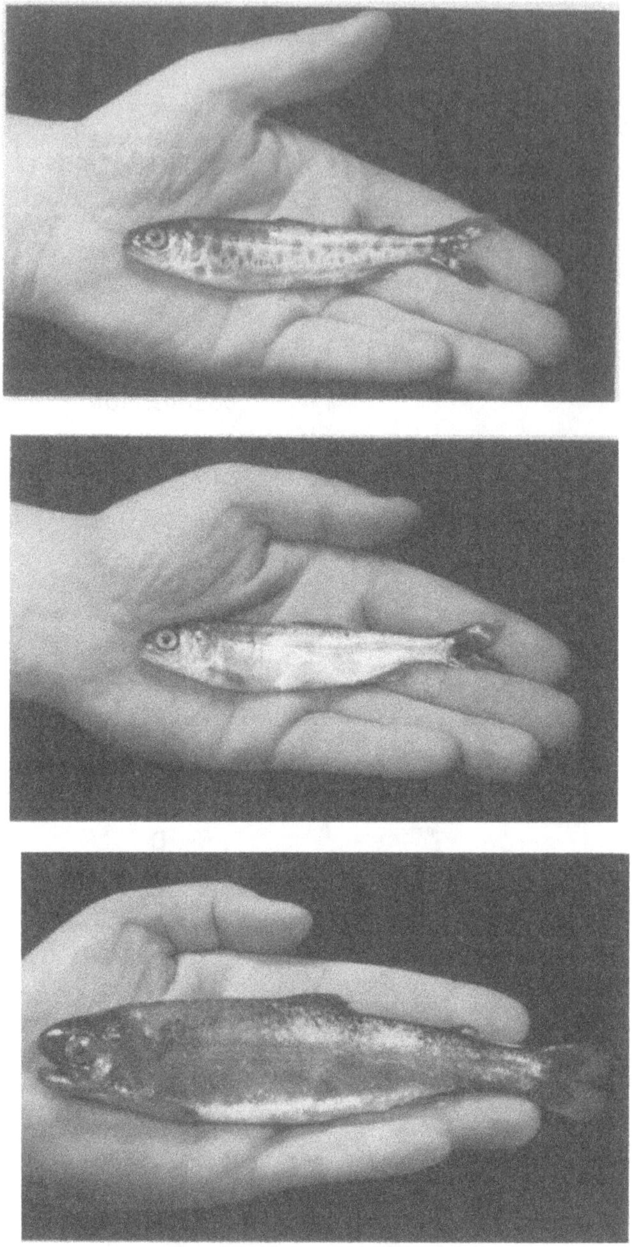

Fig. 5.5. Effect of treatment with TSH or gonadotropin on development of silvering. *Top:* fish injected with saline three times a week for 3 weeks. Note presence of distinct parr marks. *Middle:* fish injected with TSH three times a week for 3 weeks. Note the loss of parr marks and distinct silvering. *Bottom:* male salmon injected with gonadotropin three times a week for 12 weeks. Note the development of purplish-red spawning coloration and bone deformation in the snout. This fish was ripe and released milt when handled

ical and behavioral transitions in natural smolts and hormonally injected fish can be compared directly.)

These observations suggest that hormone injections accelerated smolt transformation and that the morphological, physiological, and behavioral transitions that define smolt transformation are under hormonal control. Since the physiological and behavioral transitions were not observed in saline-treated and uninjected pre-smolts, we infer that the hormones and not some other facet of the experimental procedure accelerated smolt transformation.

In summary, thyroid hormones and cortisol seem to be causal and not just correlative factors in initiating the smoltification process in the sense that: (1) their concentrations rise just before morphological behavioral, and physiological transitions occur, and (2) injection of TSH and/or ACTH into pre-smolts induces these transitions. The different physiological and behavioral transitions appear to be regulated independently by different hormones, which helps to explain why they do not always occur simultaneously in natural populations (see discussion in Chap. 3).

5.3 Experiments with Rainbow Trout

Quantification of smolt transformation is easy to perform. To illustrate this point undergraduate students in Scholz's Developmental Biology class at Eastern Washington University in the Spring Quarter of 1981 used these same procedures to quantify the effects of TSH versus saline injections in pre-smolt rainbow trout and required relatively little supervision. After 3 weeks the students found that silvering, downstream migratory behavior, salinity tolerance and osmoregulatory capability was increased in TSH-injected fish when compared to saline-injected fish.

Most investigators studying smolt transformations use the term „smolt“ loosely. Any two experts in the field may have different views about the degree of smolt transformation of an individual salmon. Since smolt transformation is a dynamic process, we argue that "smolt" is a vague term, so that seemingly conflicting results of different experimenters may not, in fact, be contradictory; instead, the explanation could center around the point that the fish used were collected in different stages of smolt transformation. Hence, we believe that quantification of the smolt transitions conveys more information than simply using the term "smolt". Since, as previously indicated (Chap. 3), there is a high degree of variability in the timing of smolt transitions in different genetic populations of salmon, quantification of smolt transitions should be done for each genetic population. In fact, we feel strongly enough about this point to suggest that journals should require a quantified definition of smolt transformation to avoid further confusion in the literature.

5.4 Methods for Studying Endocrine Control of Olfactory Imprinting

The fish used for determining the effects of TSH and ACTH on the smoltification process were also used for studying endocrine control of olfactory imprinting. The five groups in each tank were exposed simultaneously to a synthetic chemical (either morpholine at 5.7×10^{-10}M or phenethyl alcohol 4.1×10^{-8}M). For 3 weeks before receiving hormone injections, phenethyl alcohol was metered into Tank 1 and morpholine into Tank 2 (pre-treatment odor). Two days before hormone injections commenced, the odors were shut off. During the period of hormone treatment, the presentation of chemicals was switched so that Tank 1 received morpholine and Tank 2 phenethyl alcohol (treatment odor).

The hormone treatment continued for 3 weeks until the fish appeared to undergo smolt transformation, and then it was stopped at the same time the synthetic chemical was shut off. The fish were then retained for 10 months after the period of hormone treatment and odor exposure before we conducted behavioral and electrophysiological tests to determine if the fish had become imprinted to the odor present in their tank during the period of hormone administration.

Radioimmunoassays indicated that within 2 weeks after the last TSH injection the level of thyroid hormones declined to basal levels (triiodothyroxine from 12.5 to 1.0 ng ml^{-1} serum; thyroxine from 60 to 10 ng ml^{-1} serum: $n = 15$) and remained at these basal levels through the period of behavioral and physiological testing ($n = 8$).

The behavioral tests were conducted in a natural river. A meaningful response required upstream migration and selection of a tributary scented with the correct treatment odor. Electrophysiological tests were conducted by monitoring EKG's from salmon held in a flow-through test chamber when different chemicals were added to their water supply. We expected that if olfactory imprinting is connected with hormone treatment, then the fish should be able to discriminate their treatment odor in these tests. Saline-treated and uninjected fish, along with exposure to pre-treatment odor, controlled for the possibility that juvenile salmon were generally sensitive to odors independent of hormone injection. Natural smolts were exposed to morpholine or phenethyl alcohol and tested like experimentally treated fish. This permitted evaluation of responses of TSH- and/or ACTH-treated fish in comparison with "naturally imprinted" fish.

Before conducting behavioral and electrophysiological tests, half the fish from each treatment group were injected with salmon gonadotropic hormone for 12 weeks to bring them into migratory disposition and to mimic the physiological state of a naturally spawning salmon. The remainder received saline injections (controls). Sex homones, testosterone and estradiol 17-β, were monitored. In males, levels of testosterone ranged between 20 and 40 ng ml^{-1} serum in gonadotropin injected fish (n = 12) and 1 and 2 ng ml^{-1} serum in controls (n = 12). In females levels of estradiol 17-β were 5–10 ng ml^{-1} in gonadotropin-injected fish (n = 6) and 0–1 ng ml^{-1} serum in controls (n = 6). These values are comparable to those observed in adults during their spawning migration.

Gonadotropin injections were designed to mimic the normal course of events in natural spawning. Otherwise, it would have been necessary to hold the fish for an additional year before they would have spawned naturally. We believe this to be significant in the experimental procedure in view of the information presented in Chapter 4 about the effect of sex hormones on olfactory sensitivity and discrimination of home-stream odor: i.e., fish not in spawning condition would not be expected to distinguish their treatment odor. Also the nature of the field test required that the fish be in a migratory disposition before we could collect data. Salmon gonadotropin (SG-G100) for these experiments was supplied by Dr. Edward M. Donaldson, Nutrition and Applied Endocrinology Program, Canadian Fisheries and Marine Service Laboratory, Vancouver, British Columbia.

5.5 Results of Experiments on Endocrine Control of Olfactory Imprinting

There is evidence that olfactory imprinting occurred owing to TSH treatment. Pre-smolts receiving TSH or TSH plus ACTH and exposed to either morpholine or phenethyl alcohol later demonstrated the ability to track their respective odor upstream in behavioral tests, whereas subjects receiving no injection, saline or ACTH only did not. The behavior of TSH and TSH + ACTH fish was similar to that displayed by naturally imprinted fish. Electrophysiological studies of olfactory discrimination capabilities corroborate the field work. Only fish that had also received gonadotropin injections responded to their treatment odors. Details of the test procedures and results obtained are presented below.

5.5.1 Behavioral Tests

Behavioral tests were conducted by releasing fish in the Ahnapee River at Forestville, Wisconsin, 150 m below the junction of two tributaries. Morpholine and phenethyl alcohol could be introduced into either arm. The main feature of this study site is that the river is dammed by a structure with two spillways, each forming a separate tributary which become rejoined at the test area. Thus, the water forming the background for the treatment odors in both tributaries could be presumed to have uniform odor characteristics (Fig. 5.6).

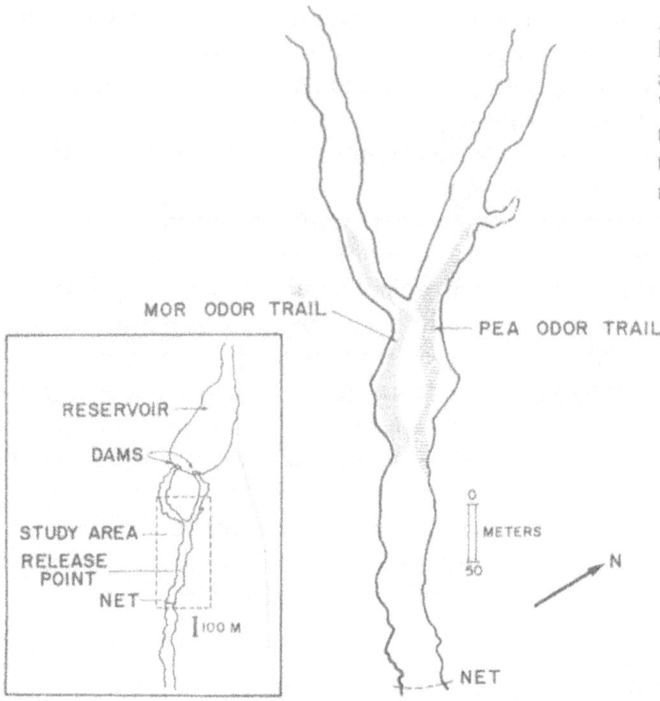

Fig. 5.6. Study site for behavioral tests on the Ahnapee River at Forestville, Wisconsin. *Inset* shows detail of how the two tributaries arise from the same reservoir

Mapping with Rhodamine B showed that below the confluence water from one tributary flowed along one shore while water from the second tributary traveled along the opposite shore (Fig. 5.6), enabling us in addition to determining which tributary the fish entered, to record their movement in relation to the odor trails. If their odor were present, we expected that fish would swim upstream on the side of the river scented with their treatment odor, following the edge of the odor tail, and select the tributary from which the odor was emanating. If their odor was absent, we expected fish to swim downstream. A net strung across the river about 500 m below the release point prevented the fish from migrating farther downstream.

Our expectations were derived from the study by Johnson (1978), who observed that sexually mature adult coho salmon exposed to morpholine during the smolt stage migrated upstream if morpholine was present in the stream and downstream if morpholine was absent. He concluded that the odor acts as a sign stimulus to release a stereotyped behavior pattern, i.e., swimming against the current, and suggested that this is the operational mechanism in selection of the home stream. Johnson also introduced morpholine on the left or right side of the stream and observed by dye mapping that the odor trail did not extend beyond midstream. He found that the fish movements were confined to the right half of the stream when morpholine was introduced on the right side and to the left half when it was introduced on the left side. The fish did not swim within the odor trail but instead along the edge, thereby reducing the possibility of sensory adaptation to the odor.

In the present experiment, each fish was given three trials: two with its correct odor present in either arm and one with the odor absent. For at least two trials their pre-treatment odor was also present, thus producing an element of choice in this test. Therefore, three criteria had to be fulfilled for the behavior of experimental fish to be equivalent to that of natural adults toward their imprinting odor:

1. Upstream migration if the treatment odor was present along the edge of the odor trail and on the side of the river scented with the odor.
2. Selection of the tributary scented with correct odor.
3. Downstream migration if the odor was absent.

Fish meeting these criteria were classified as "imprinted" and those not fulfilling them were classified as "not imprinted".

Fish were identified by color-coded floy tags attached to the base of the dorsal fin and tracked visually. A "double-blind" procedure was used for making observations of fish behavior. At least two observers were present for all tests. Neither observer knew the treatment of the fish. Only one of the observers added the odor to the tributaries, so the second observer did not even know which odor was being tested. A steady-state concentration of 5.7×10^{-10}M morpholine or 4.1×10^{-8}M phenethyl alcohol was maintained in the test area throughout the tracking period.

Our results (Fig. 5.7, Table 5.1) indicated that TSH injections induced olfactory imprinting in pre-smolt coho salmon. Pre-smolts receiving TSH and simultaneously exposed to synthetic chemicals later demonstrated the ability to track their treatment odor upstream, whereas those receiving ACTH, saline, or no injections did not. The behavior of TSH-treated fish was similar to that displayed by naturally imprinted fish. Therefore, TSH injections mimic the events that activate olfactory imprinting in natural smolts.

Only gonadotropin-injected hatchery smolts, treated with TSH or TSH+ACTH displayed this attraction. Fish not receiving gonadotropin were not in a migratory state and rarely moved from the vicinity of release. In holding position underneath overhanging banks, in pools, or by submerged trees, their behavior was not unlike that of parr. In contrast, fish receiving gonadotropin were in a migratory state: hatchery smolts, TSH, and TSH+ACTH fish swam upstream if their odor was present and downstream when it was absent. ACTH-, saline-treated and uninjected fish invariably swam downstream when their treatment odor was present or absent.

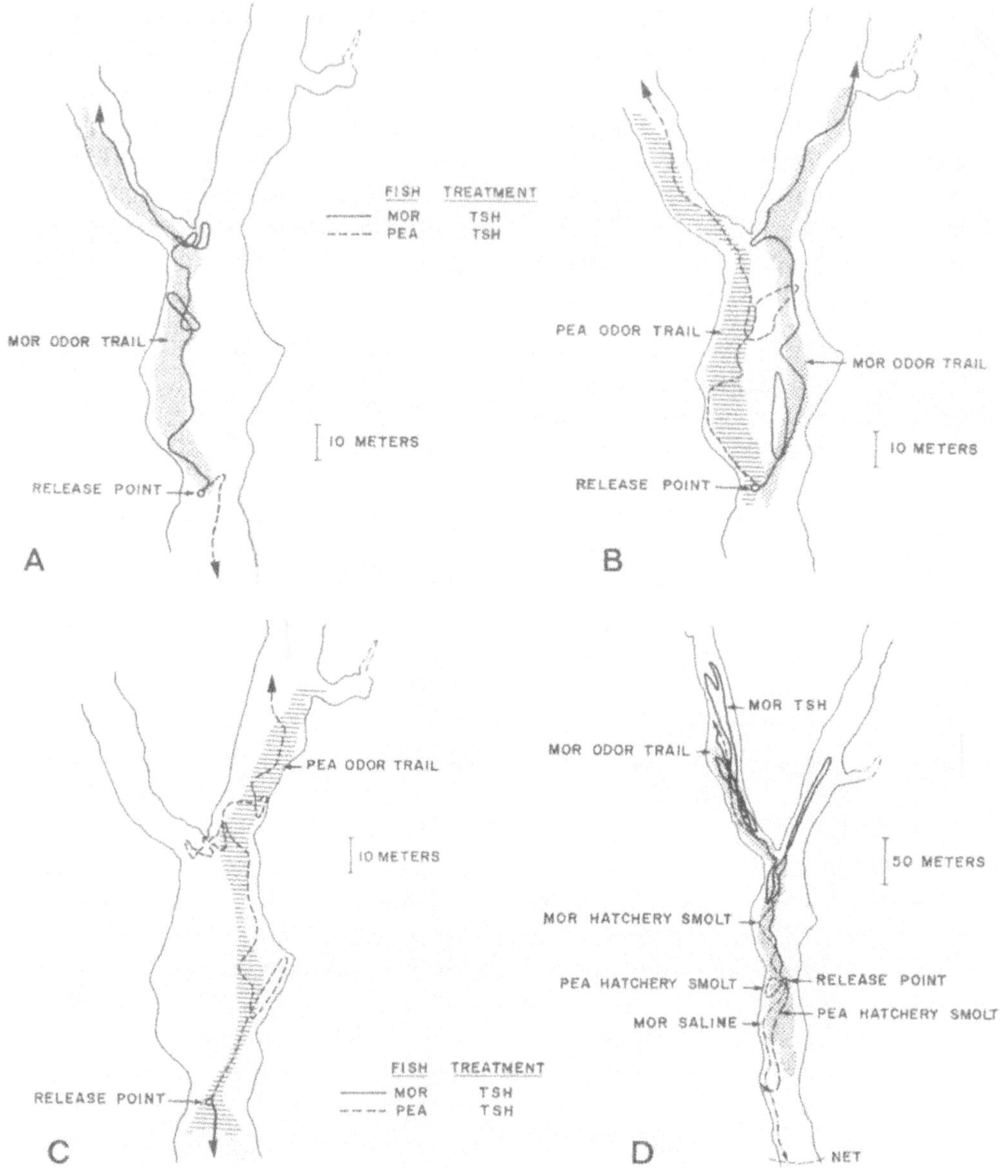

Fig. 5.7 A–D. Representative plots of fish tracked in field behavior tests. Note that fish travel predominantly on the side of the river scented with their treatment odor, along the edge of the odor trail, before selecting the appropriate tributary. In *A–C* fish in each diagram are the same two (TSH-injected fish exposed to either morpholine or phenethyl alcohol) fish. *D* shows typical plots of fish with other treatments. Note that the behavior of TSH-injected fish resembles that of natural (hatchery) smolts. Note in *D*, that TSH-injected fish initially selected the tributary without the treatment odor and then back-tracked. Morpholine-exposed fish receiving saline instead of TSH did not respond to their treatment odor even if they had been injected with gonadotropin

Table 5.1. Summary of results from behavioral tests. Fish grouped by treatment

Group	Treatment	Odor	n=	Odor absent (n x 1 trial) Down	None	Up	Odor present (n x 2 trials) Down	None	Up	Odor discrimination MOR	PEA
G+	SMOLT	MOR	10	9	1		1	1	18	18	
		PEA	8	8			1		15	1	14
	TSH + ACTH	MOR	13	11	1	1	2		24	21	3
		PEA	12	11		1	2		24	2	20
	TSH	MOR	8	8			1		15	14	1
		PEA	8	6	1	1	1	1	14	2	13
	ACTH	MOR	8	7	1		15		1		
		PEA	4	4			7		1		
	SALINE	MOR	13	12		1	27	1	1		
		PEA	11	9	1	1	20		2		
	UNINJECTED	MOR	6	6		1	11		1		
		PEA	4	3			8				
SALINE +	SMOLT	MOR	6	1	4	1		10	2		
		PEA	4		3	1		7	1		
	TSH + ACTH	MOR	6		5	1	1	9	2		
		PEA	4		4		1	6	1		
	TSH	MOR	6		5	1	1	9	2		
		PEA	3	1	1	1		3			
	ACTH	MOR	4		4			7	1		
		PEA	0								
	SALINE	MOR	6	1	5		2	8	2		
		PEA	4		2	2	2	5	1		
	UNINJECTED	MOR	4		3	1	1	6	1		
		PEA	0								

5.5.2 Electrophysiological Tests

Electrophysiological tests were conducted by monitoring heart rate (EKG) of TSH- and saline-injected salmon which were held in an experimental chamber and presented with morpholine or phenethyl alcohol (Fig. 5.8). A reduction in heart rate within 1–4 heartbeats following presentation of an odor indicated that a fish could detect it. A response to the treatment odor but not the alternate odor would indicate that the fish had become imprinted to the treatment odor and could distinguish it from other odors. Hirsch (1977, pers. commun.) had previously shown that sexually mature adult salmon which had been imprinted to either morpholine or phenethyl alcohol during the smolt stage usually responded to their imprinting odor but rarely to other odors. Thus, the major criterion of a response is that the fish show preferential reactions to a particular odor. The EKG tests were performed by Mark Muzi, who did not participate in the field behavior test. Hence, the electrophysiological tests provided an independent measure of the effect of hormone administration upon imprinting.

Fish were placed in a covered box and various odors were presented to them in their water supply (Fig. 5.8). The odor signal is (1) transduced into

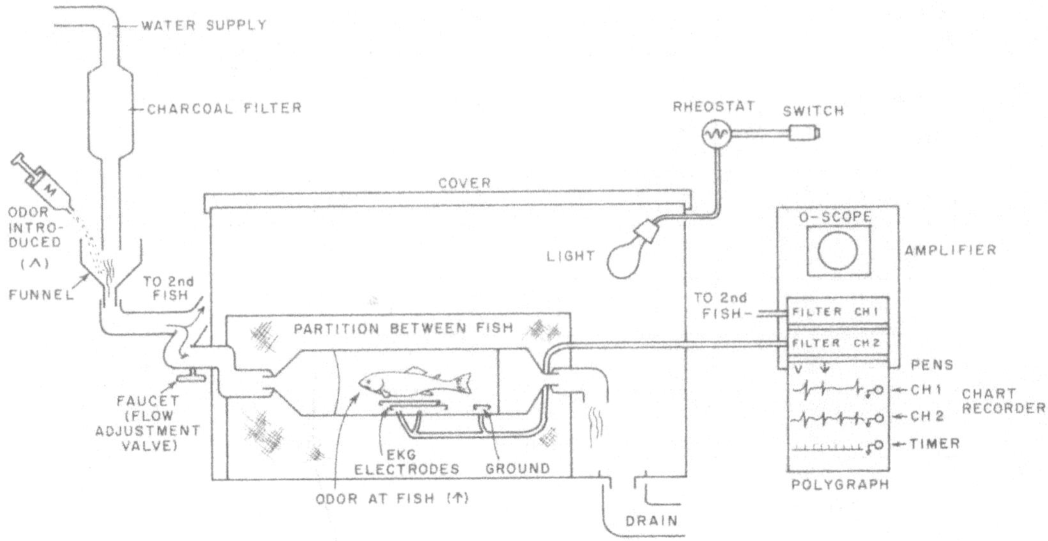

Fig. 5.8. EKG apparatus used for physiological tests of olfactory discrimination

a series of nerve impulses in the olfactory epithelium, (2) integrated in the fish's brain, and (3) transmitted by parasympathetic cholinergic fibers in the vagus nerve to the heart. Release of acetylcholine at the pacemaker region from these nerves reduces the rate of depolarization of the pacemaker potential, thereby increasing the interval between action potentials or the time between successive heartbeats, so heart rate is decreased. The size and speed of the myocardial movements are detected by exterior carbon electrocardiogram (EKG) electrodes. The signals received by the electrodes are amplified and visually displayed on an oscilloscope or a chart recorder (Grass model 790 recorder with a 705–13 preamplifier, a 7DAE amplifier, and a model H-25-60 chart drive recorder). A Faraday cage suspended over the experimental apparatus shielded the heart-rate signal from 60 Hz noise.

Test subjects were put into acrylic tubes closed at each end with a Büchner funnel so that water flowed through the tubes in one direction. The tubes had EKG electrodes mounted in them and were encased in the covered box.

A continuous flow (1–2 l min^{-1}) of aerated, charcoal-filtered tap water (Madison City supply) was delivered to the fish by way of PVC tubing connected to the acrylic tubes through holes in the box, where a pulse of odor could be introduced via glass syringe. This arrangement allowed us to introduce odors without the fish viewing the procedure.

The time required for an odor to reach the fish through the water supply was determined visually by presentation of methylene blue dye into the funnel before and after fish were tested with various odors. Dilution factors of the odors were also calculated from this flow. Odors tested included:

1. Morpholine (5.7×10^{-10}M).
2. Phenethyl alcohol (4.1×10^{-8}M).
3. Blank syringes filled with the background water to control for the possibility that responses could be elicited due to experimental procedure independently of the odor cues.
4. Light on/off – a 30-W light mounted in the experimental chamber was either turned on or extinguished abruptly. This normally results in an immediate deceleration in HR. This test controlled for the general sensitivity of the fish. The response was immediate since the stimulus was detected as soon as it was presented, unlike stimuli which travel through the water supply. We reasoned that if the fish failed to respond to this major alteration in the environment, then something was wrong with the fish; so if no response was observed, the fish was eliminated from the data set.

Fish were placed in the experimental chamber and acclimated for 24 h before odor discrimination tests were conducted. Each fish was tested in five to ten blocks with two trials for each odor, blank, and light on/off presented in random order per block. Successive blocks were separated by a 20-min to 12-h waiting period.

Quantification of response in heart rate was accomplished by measuring the distance between each of 25 successive heartbeats prior to presentation of the odor and calculating mean and standard deviation. Since all of the heartbeats were within three standard deviations, it seemed feasible that a heartbeat outside this range could be considered a response. The first four heartbeats after arrival of the odor at the fish were measured. If one or two of these heartbeats was greater than the average duration +3 S.D., we recorded that the fish responded to the odor by a heart-rate deceleration. The number of responses of each fish to a particular odor was divided by the total number of trials in which the odor was presented. Thus, the percent responsiveness of each fish to each odor could be determined. Statistical comparisons were made among groups of fish with different hormone treatments and odor experience by using the Mann-Whitney Rank test. In practice, determination of a response could be made visually in most cases (Fig. 5.9). Before an odor stimulus was introduced heart rate was generally very regular and deceleration was pronounced if a response to the odor occurred.

The types and numbers of fish and results are recorded in Table 5.2. TSH-injected fish discriminated their respective treatment odors by a reduction in heart rate, whereas saline-injected fish did not. Only TSH-treated fish which had been treated with gonadotropin displayed this

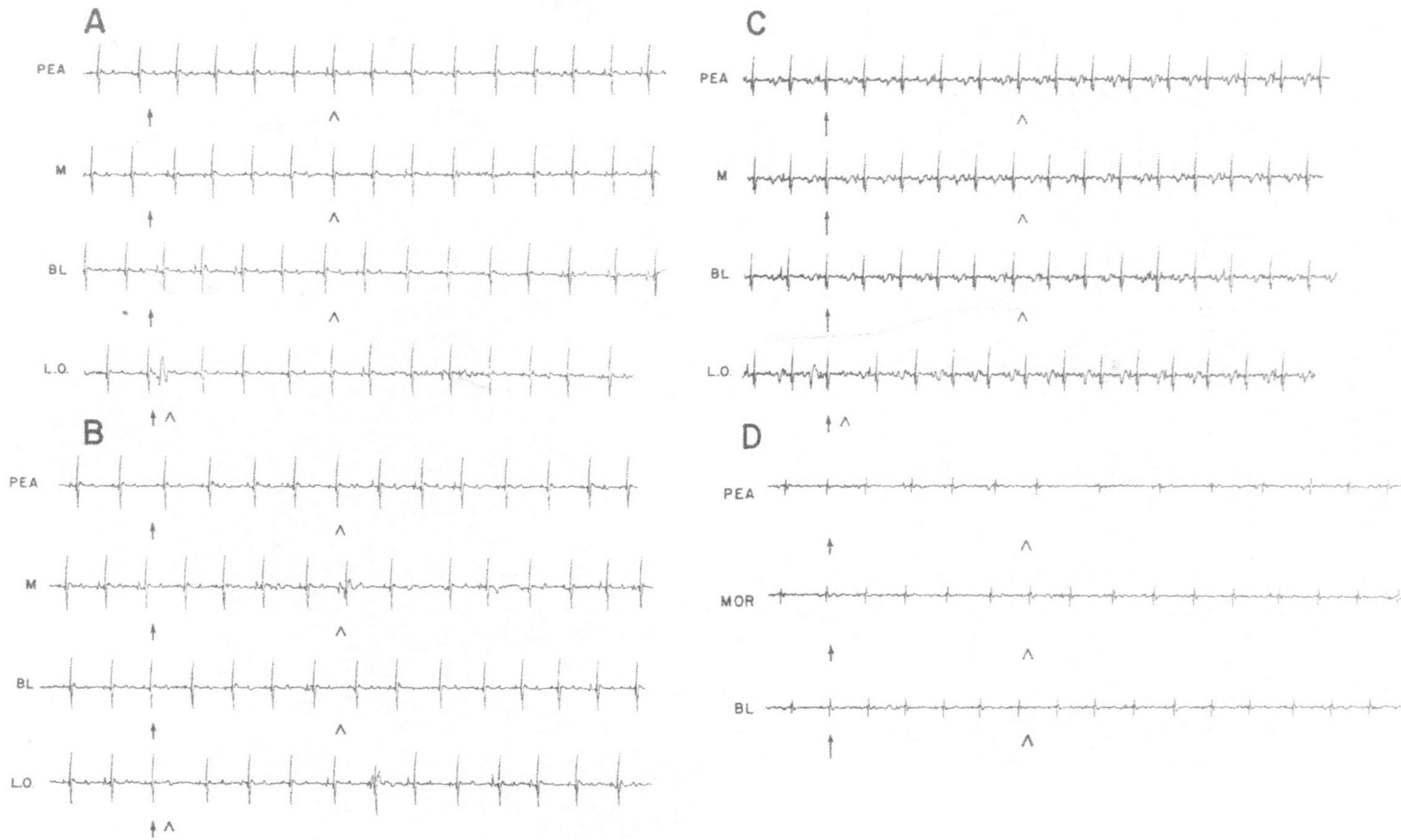

Fig. 5.9A–D. EKG traces of *A* saline-injected (control) salmon exposed to morpholine (no HR deceleration to morpholine); *B* TSH-injected salmon exposed to morpholine (note that HR deceleration occurs after morpholine but not PEA or blank arrives at the fish); *C* saline-injected fish exposed to PEA; *D* TSH-injected fish exposed to PEA (note that HR deceleration occurs after PEA but not morpholine or blank arrives at the fish). In *A–D* the odor presented is indicated on the *left* of the EKG trace. *PEA* phenethyl alcohol, *M* morpholine, *BL* blank, *L.O.* light. Note the deceleration in heart rate to the morpholine but not PEA or BL. ↑ stimulus introduced; ∧ arrival at fish. All fish had been injected with gonadotropin prior to testing

Table 5.2. Summary of EKG of individual salmon arranged by treatment group

Fish no.	Clip	Hormone treatment	Odor	GTH (Y/N)	MOR		PEA		Control		L.O.	
					n=	%	n=	%	n=	%	n=	%
100	ALP	TSH	MOR	Yes	22	100	19	5	22	0	17	82
101					8	88	5	20	5	0	4	75
102					12	67	16	0	12	8	6	83
103					10	30	7	14	5	0	4	50
104	LP	Saline	MOR	Yes	22	0	22	0	20	0	20	100
105					12	8	14	8	11	0	15	80
106					10	0	16	0	14	0	13	77
107	ARP	TSH	PEA	Yes	20	0	16	88	16	0	12	100
108					15	0	17	59	13	0	12	100
109					15	20	17	94	17	5	18	67
110					13	15	12	92	13	8	17	86
111					12	0	13	100	12	8	16	77
112	RP	Saline	PEA	Yes	19	0	19	0	21	0	10	80
113					13	15	21	0	10	0	16	100
114					14	7	15	7	11	0	14	79
200	ALP	TSH	MOR	No	14	43	18	38	24	8	19	78
201					21	14	24	17	25	0	20	95
202					26	15	24	17	21	4	18	88
203					20	5	17	6	16	0	13	92
204	LP	Saline	MOR	No	21	14	20	25	21	0	20	95
205					19	22	18	50	19	0	10	100
206					4	50	4	25	9	0	7	71
207					4	50	4	50	9	0	7	71
208					10	20	9	11	10	0	10	100
209	ARP	TSH	PEA	No	22	0	20	5	17	0	20	80
210					11	18	11	5	15	0	11	100
211					13	8	15	15	15	0	8	25
212					16	13	13	13	13	0	6	83
213					5	40	6	33	4	0	–	–
214	RP	Saline	PEA	No	8	50	8	13	13	0	7	100
215					19	11	18	11	21	0	20	80
216					2	50	2	50	4	0	–	–
217					7	14	6	33	8	0	–	–
218					7	43	6	17	8	0	–	–
219					14	21	8	13	14	8	9	67

MOR = morpholine, PEA = phenethyl alcohol, blank = control, L.O. = light on/off.
Dash indicates that test was not administered

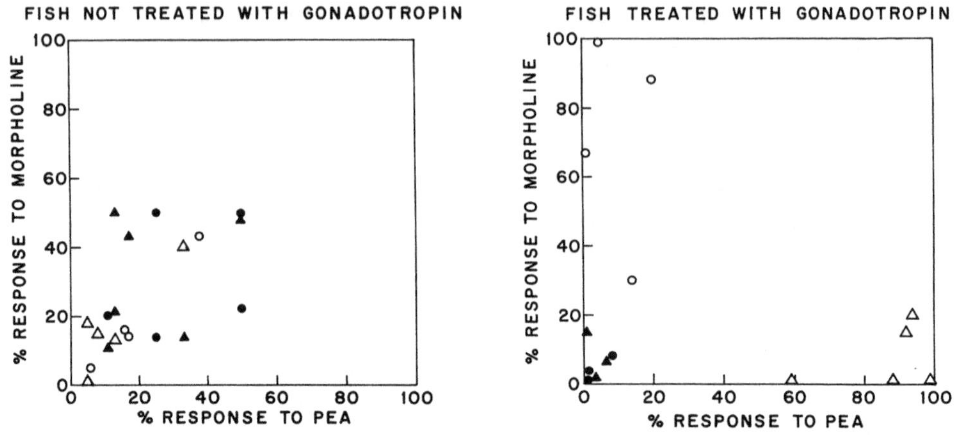

Fig. 5.10. Summary of EKG responses in fish not treated with gonadotropin. Responses to morpholine plotted against responses to phenethyl alcohol for each experimental group. % response = no. of trails tested divided by no. trials responded to the odor × 100. Initial treatment: △ phenethyl alcohol, TSH; ▲ phenethyl alcohol, saline; ○ morpholine, TSH; ● morpholine, saline

response. Fish rarely responded to blank samples, indicating that the odors and not the experimental procedure produced these results.

Gonadotropin injection increased the olfactory sensitivity of salmon to their imprinting odor while actually decreasing their sensitivity to other odors. In the EKG tests, TSH-injected fish not receiving gonadotropin responded indiscriminately 20%–30% of the time to both morpholine and phenethyl alcohol, whereas those receiving gonadotropin responded 85–100% of the time to their treatment odor and 0%–10% of the time to the alternate (Fig. 5.10). Saline-injected fish not receiving gonadotropin responded to both odors (20%–35%) while those receiving gonadotropin responded to neither (0%–10%). The differences between the gonadotropin-treated fish and their saline counterparts were significantly different (Mann-Whitney U Rank test, Siegel 1965: $U = 15$; $P \leqq .01$). These results imply that during the spawning migration, gonadotropin or sex hormones act on the CNS to exert centrifugal control over the olfactory system, perhaps either by desensitizing the receptors to all but the imprinted odor, or altering central integration processes.

5.6 Organizational Influence of Thyroid Hormones on Maturation of the Central Nervous System

The external indicators most closely associated with imprinting are silvering and downstream migratory behavior because, like the imprinting process, these transitions seem to be regulated by thyroid hormones. Development of salinity tolerance and osmoregulatory capability is not a good indicator of imprinting because these transitions seem to be controlled by ACTH through cortisol, which does not influence the imprinting process.

The underlying mechanisms by which thyroid hormones activate olfactory imprinting were not delineated in the work cited above. The range of possibilities of thyroid influences could include factors as simple as increased olfactory sensitivity to odors, or complex processes like development of synaptic connections in the central nervous system. However, any mechanism must take into account the facts that (1) formation of a permanent "long-term" olfactory memory resulted from TSH treatment (or high thyroid hormone levels); i.e., fish responded in a stereotyped pattern of behavior to their respective treatment odor 10 months after being exposed to it, and without again encountering the odor during the intervening period; and (2) shortly after the period of TSH injection thyroid hormone levels returned to normal, so this memory persisted in the absence of elevated levels of thyroid hormones. This would seem to rule out many of the simple factors that could account for the action of thyroid hormones, and points to an organizational influence such as thyroid induction of structural changes, e.g., increased arborization and formation of synaptic connections during the imprinting process that would survive the later decreases in concentration of thyroid hormones. The basic mode of operation must be different for thyroid involvement in downstream migratory behavior of smolts which, because downstream migratory activity ceases when thyroid hormone levels recede, is an activational influence. We suspect that imprinting in salmon requires an alteration in neural substrate and that this alteration requires an increase in the concentration, and perhaps in the binding of thyroid hormones in the brain at a critical period of development corresponding with smolt transformation. This hypothesis is supported by the association of elevated thyroid hormone levels with critical periods for brain maturation and with continued mental function in a variety of vertebrates.

Thyroid hormones have well-documented organizational influences on maturation of the central nervous system in vertebrates (reviewed by Hamburgh 1968, 1969). Work with neonatal rats suggests that thyroid hormones have the property of influencing maturation of the CNS during a "critical" or sensitive period of development. Several lines of evidence have established that the neonatal rat brain has an obligatory requirement for thyroid hormones within 10–14 days after birth. Natural or experimentally induced (by radiothyroidectomy) hypothyroidism during the "critical" or "sensitive" period results in abnormal differentiation in brain-cell structure, as evidenced by deficient arborization of neuronal processes and retardation in synapse formation as determined by the rapid Golgi stain method (Legrand 1967a, b, 1979, Lander 1979, reviewed by Hamburgh 1968, 1969, Rabie et al. 1979). Thyroid hormones, TSH, or TRF administered to hypothyroid rats during the critical period, but not at later times, resulted in normal arborization of these dendrites (reviewed by Hamburgh 1968, 1969). Defects in brain maturation can be only minimally restored if thyroid-hormone replacement is started after the critical period. These data clearly indicate a role for thyroid hormones in facilitating the development of the framework requisite to normal neural function.

Additionally, neonatal hypothyroidism in rats has been correlated with impairment of certain types of behavior and learning deficits (Eayrs 1964, 1966, Essman et al. 1968, Schalock et al. 1977) and hyperthyroidism with precociousness in motor and learned responses (Schapiro 1968, Davenport and Gonzales 1973). In the hypothyroid animals, injections of thyroid hormones during the critical period, but not at later times, result in normal behavior and learning.

In view of the temporal correlation between thyroid influences on maturation of the CNS and thyroid influences on behavior and learning, Eayrs (1966) has speculated that: (1) abnormal development of the structural architecture of the neuron can be considered responsible for the permanent impairment in learning ability, and (2) organizational influences of thyroid hormones on behavior and learning are related to formation of synaptic connections in the CNS.

Hamburgh (1969) pointed out that to establish thyroid influences on maturation of the central nervous system, it must first be demonstrated that the areas undergoing differentiation are targets of thyroid hormones. Evidence for brain sensitivity to thyroid hormones during the critical period for CNS development is shown in recent studies with juvenile rats (Naidoo et al. 1978, Schwartz and Oppenheimer 1978, Dozin-van Roye and De Nayer 1979, Ishiguru et al. 1980). In these studies thyroid hormone receptors were characterized in a developmental sequence from birth through day 55. Affinity constants (Ka) and maximum binding capacity (R_0) were derived from Scatchard plots. Binding affinity increased during the critical period between day 10 and 14 and then declined. This indicates that maximum thyroid hormone receptor affinity correlates well with the period of morphological changes for the cerebrum (Dozin-van Roye and De Nayer 1979).

In the metamorphic tadpole a similar surge of thyroid hormones (Mondon and Kaltenbach 1979) occurs concommitantly with brain maturation (Gona 1974, Gona and Gona 1977). Amphibian metamorphosis is a developmental process similar to smoltification in the salmon. Therefore, we postulate that the elevated concentrations of thyroid hormones experienced by the smolting fish could also result in maturational brain changes.

This developmental role for thyroid hormones may also be involved in the imprinting process in young birds. While this correlation has never been tested, several independent studies, taken together, suggest this possibility. Hess (1973) has shown that the period for imprinting begins while the young chick is in the egg and extends until after it hatches. Thommes et al. (1977) have measured plasma levels of thyroid hormones in the pre- and post-hatching chick; their results show that a peak in the thyroid hormone levels occurs at hatching. Finally, thyroid-blocking experiments with pre-hatching chicks demonstrated evidence of delayed maturation of the central nervous system (Chandrasekhar et al. 1979) similar to that discussed for the post-natal rat. While these studies do not definitely link elevated levels of thyroid hormones with both imprinting and brain maturation, the developmental proximity of these events suggests a correlation.

Currently, at Eastern Washington University (Scholz and Ronald J. White) and at the University of Wisconsin (Marion Meyer) we are testing the hypothesis that the surge of thyroid hormones during smolt transformation in coho salmon (1) activates the olfactory imprinting process, and (2) influences maturation of the brain. The latter will be determined by (a) localization and characterization of receptors for thyroid hormones in brain tissue by Scatchard plot analysis to determine binding affinity (K_a) and receptor concentration (R_0) and, (b) neurohistological staining of brain regions that contain high-density, high-affinity receptors.

Our basic procedure is to monitor the surge of thyroid hormones in salmon undergoing natural smoltification, or to manipulate experimentally the thyroid hormone levels out of phase with natural cycles by either injecting TSH into pre-smolts to accelerate the thyroid surge, or blocking the thyroid surge with thiouracil or low iodine diet, and in each case, measuring the effect upon olfactory imprinting and brain-neuron maturation.

In our study we are really measuring the effect of thyroid hormone on two independent events, olfactory imprinting and brain maturation. Correlations between the thyroid–olfactory imprinting interaction and thyroid–brain maturation do not imply a cause and effect relationship between brain maturation and the imprinting process. However, if we obtain data indicating that olfactory imprinting and brain maturation occur simultaneously during natural development, and if both events occur simultaneously out of phase with the normal developmental sequence as a result of hormone manipulation, we could strengthen the hypothesis that there is a mechanistic relationship between the two events.

We believe that our imprinting model for salmon offers distinct advantages over other systems for exploring brain–thyroid interactions associated with learning and memory. In mammals, birds, and amphibians, the critical period occurs early in development; therefore, specific thyroid influences upon learning and memory cannot be separated from generalized brain maturation. In fact, all parts of the brain may be subject to thyroid-induced maturation (Ford and Crammer 1977, Patel et al. 1979, Rabie et al. 1979, Valcana 1979). However, since smolt transformation occurs relatively later in development, after the period of generalized brain maturation, the effects of thyroid hormones might not be so pervasive in the salmon and only a few brain areas associated with imprinting might be affected by thyroid hormones.

Chapter 6

Endogenous and Environmental Control of Smolt Transformation

This chapter considers endogenous and environmental factors that influence smolt transformation. It also provides evidence that an endogenous developmental program, mediated by hormones and entrained by the annual cycle in daylength, regulates the smoltification process.

6.1 Size

Fish that reach a certain critical size undergo at least some of the physiological and behavioral transitions under controlled laboratory conditions (Hoar 1976). In the field a certain minimum size must be achieved for smolt transformation in Atlantic, sockeye, and coho salmon. For coho salmon the critical size of aprroximately 15 cm is normally reached by 1.5 years of age (Drucker 1972). Because of the slower growth rate the critical size is achieved later in colder, less productive regions (Drucker 1972). Accordingly, smolt transformation occurs at a later age. However, size does not appear to be the only factor that governs the timing of smolt transformation, since some parr may reach the "critical" size 6 months before the smoltification process occurs. Because smolt transformation normally occurs in the spring, these observations imply that a fish must reach a certain critical size by the end of winter in order to be a candidate for smolt transformation in the oncoming spring. Experimental evidence suggests that (1) environmental cues, including daylength and water temperature, act as external regulators to synchronize the precise timing of smolt transformation, and (2) these cues activate the neuroendocrine system which serves as a link between the fish and its environment (see following sections on water temperature and photoperiod).

6.2 Water Temperature

One factor that fluctuates on a seasonal basis and could, therefore, serve as a cue for initiating smolt transformation is water temperature. Influence of increasing water temperature on smolt transformation has been studied in the field and under controlled laboratory conditions.

Several field studies have reported that the time of the downstream migration of smolts is affected by water temperature (reviewed by Thorpe and Morgan 1978). One way of studying this relationship in sockeye, coho and Atlantic salmon, and in rainbow trout has been to record the number of downstream migrants captured in traps and compare them with fluctuations in water temperature (Hayes 1953, Hartman et al. 1962, Northcote 1962, Osterdahl 1969, Solomon 1975, Munro, unpublished). In these studies migration was initiated after the water temperature had increased to about 10 °C. Once the migration was underway, sudden decreases in water temperature were accompanied by reduced numbers of fish caught in the traps. Afterward, as the temperature increased, more fish were caught. In addition, some ultrasonic tracking studies with Atlantic salmon smolts suggest that the smolt migration begins when water temperature reaches 9–10 °C: in years when the water warms up quickly, the peak of migratory activity occurs earlier than in years when the water warms up more slowly (Fried et al. 1978, LaBar et al. 1978, McCleave 1978, Solomon 1978, Tytler et al. 1978).

One problem with these field studies is that temperature cues do not function independently but, in fact, fluctuate in concert with daylength, making the individual effects of these environmental variables on smoltification difficult to separate. For example, in one of the ultrasonic tracking studies mentioned above, McCleave (1978) made observations on two successive year classes of salmon smolts from one river. He reported that while peaks in downstream migratory activity did not occur until after the temperature had reached 10 °C (on different dates each year), the date that the first fish left the river in both years was May 10.

Examination of data for several years of trapping from a single river suggests that the date on which the first fish leave the river is fairly constant from year to year, but that there is substantial variation in water temperature at that date (Elson 1962, Foerster 1968, Bakshtansky et al. 1976a, b). Thus, it seems unlikely that temperature is a causative factor in onset of downstream migration, i.e., the timing of the migration does not deviate from calendar date as much as would be expected if it were related only to climatic conditions.

This idea is corroborated by laboratory studies in which temperature and photoperiod could be controlled independently. In one study, the intensity of rheotactic responses (swimming upstream against current) of juvenile Pacific salmon declined with increasing temperature (Keenleyside nd Hoar 1954). However, in most laboratory studies, short-term increases (from 6° to 12 °C) in temperature with photoperiod held constant, designed to simulate natural spring temperature, produced no marked silvering, salinity tolerance, or migratory behavior in coho salmon or rainbow trout (Johnston and Eales 1968, Saunders and Henderson 1970, Wagner 1974b).

Long-term increases in temperature, for example, raising or acclimating fish at 12 °C, had little effect on silvering or migratory activity but did increase salinity tolerance and gill Na^+/K^+-ATPase activity (Zaugg and McLain 1972, Adams et al. 1973, 1975, Wagner 1974b). Increasing the acclimation temperature to 15 °C results in a reduction of salinity tolerance and Na^+/K^+-ATPase. Since N^+/K^+-ATPase from salmon gills appears to be temperature-sensitive, it has been suggested that the enzyme has an optimal efficiency at a temperature of about 12 °C in coho salmon.

6.3 Daylength

The timing of smolt transformation is predictable and regular. It generally occurs at about the same time annually in spring in different populations of salmon at a given latitude. In Wisconsin, Washington and Southern British Columbia, for example, it happens in late April or early May. Consequently, lengthening of the days in the spring probably serves as an environmental cue in synchronizing the smoltification process. Photoperiodic cueing of smolt transformation is an attractive notion because the daylength on a given date is constant from year to year and salmon smolt on approximately the same date every year (reviewed by Hoar 1976, Poston 1978, Scholz 1980). In addition, artificial increase of daylength several months before smolt transformation would normally occur stimulates many of the behavioral and physiological transitions usually associated with smoltification. In contrast, in cases in which the normal vernal increase in daylength was experimentally delayed, fish did not smolt at the correct time, but instead, at a date later than usual (reviewed by Hoar 1976, Bern 1978, Clarke et al. 1978, 1980, Poston 1978).

Increasing daylength artifically several months before the normal schedule causes early smolt transformation in several species of salmonids. In this type of experiments, fish were reared under artificial lights so that either their photoperiod was advanced 3 months ahead of the natural photocycle, or in some of the experiments the annual photoperiod cycle was accelerated incrementally by adjusting both phase and frequency. Fish raised under both regimes underwent smolt transformation significantly earlier than control fish maintained under normal daylength (in January or February and late April to early May respectively; reviewed by Hoar 1976, Poston 1978). Physiological and behavioral transitions specifically affected by early exposure to a long daylength include:

1. Salinity tolerance and preference. In some studies with coho, Atlantic and sockeye salmon, and rainbow trout artifically increasing day lengths in January caused an earlier development of salinity tolerance (Bagger-

man 1960 b, Komourdjian et al. 1976 a, b, Bern 1978, Clarke et al. 1978, 1980), increased osmoregulatory capability (Conte et al. 1966), and increased Na^+/K^+-ATPase activity in the gills (Ewing et al. 1979). However, other studies indicate that salt-water adaptation and Na^+/K^+-ATPase activity in rainbow trout appeared not to be affected by manipulation of daylength (Zaugg and Wagner 1973, Wagner 1974 a, b).
2. Silvering. Advancing or accelerating changes in daylength caused disappearance of parr marks and silvering in coho and Atlantic salmon and rainbow trout (Hoar 1965, Johnston and Eales 1968, Saunders and Henderson 1970, Wagner 1974 b).
3. Downstream migratory behavior. Rainbow trout exposed to accelerated change in daylength began migrating downstream earlier than controls exposed to a normal daylength (Northcote 1958, Zaugg and Wagner 1973, Wagner 1974 b). Migratory tendencies in these studies were assessed by releasing the fish into natural streams and subsequently trapping and enumerating the downstream migrants.

Delaying the normal increase in daylength at the time that smolt transformation normally occurs retards the smoltification process. Physiological and behavioral transitions that are affected include:

1. Salinity tolerance or preference. Exposure of coho salmon to constant short days (8L:16D) inhibits the normal increase in salinity tolerance and salt-water preference (Baggerman 1960 b, Zaugg and Wagner 1973). Na^+/K^+-ATPase activity in gills is also not as high in experimental fish as in controls (Zaugg and Wagner 1973).
2. Coloration. Silvering is suppressed in rainbow trout and coho salmon held under constant short days (8L:16D) (Baggerman 1960 b, Van Overbeeke, cited in Hoar 1965).
3. Downstream migratory behavior. Seaward migration of rainbow trout, coho or Atlantic salmon held in the dark or under constant daylength was impeded compared with that of controls held under natural daylength. In the control groups the peak of the downstream migration occurred at the normal time in late April to early May, whereas in the experimental groups, it was delayed until July (Wagner 1974 a, b, Isaksson 1976).

6.4 Endocrine Mediation of Photoperiodic Cues

Evidence that the influence of daylength on smolt transformation is mediated by hormones is fragmentary.

6.4.1 Thyroid Hormones

Histological indications of increased thyroid activity have been reported in coho salmon raised in accelerated or advanced photocycles (Baggerman 1960 b, reviewed by Hoar 1976, Poston 1978).

6.4.2 Osmoregulatory Hormones

Prolactin. Histological evidence and electrophoretic determinations suggest that PRL levels decline in salmon subjected to increasing daylengths (Komourdjian 1976).

ACTH, Cortisol. No reports are available on the influence of daylength on secretion of cortisol.

Growth Hormone. GH production in Atlantic salmon is increased at an earlier date in experimental fish exposed to an accelerated increase in daylength photoperiod compared with control fish exposed to natural daylengths (Komourdjian et al. 1976 a, b, Bern 1978, Clarke et al. 1980).

6.4.3 Time-Keeping Role of the Pineal Gland

A time-keeping role has been proposed for the pineal gland in salmon (reviewed by Poston 1978). Changes in daylength detected and coded by the pineal gland are conducted via nerves to the hypothalamus where they are converted into neuroendocrine signals that act as inhibiting or releasing factors for pituitary hormones.

6.5 Endogenous Factors Regulating Smolt Transformation

The consensus of studies on effects of daylength on smoltification is that factors responsible for smolt transformation are completely internal since fish, even when held in darkness at a constant temperature, eventually smolt. Wagner (1974a, b), for example, reported that photoperiod manipulations could not permanently block smolt transformation in rainbow trout because fish that were reared entirely in darkness at a constant temperature (6 °C) from the egg stage eventually displayed downstream migratory activity. He concluded that smolt transformation is regulated by an endogenous rhythm which can be readjusted by external factors, e.g., daylength. The ecological importance of photoperiodic readjustment of endogenous cycles in salmon may be that it functions to trigger emigration while water levels are high in spring. This would make passage downstream easier than in midsummer.

6.6 Influence of Daylength on the Smoltification Process and the Production of Thyroid Hormones

We have conducted research on mechanisms through which daylength influences the salmon endocrine system and regulates the precise timing of smoltification. In these experiments coho salmon, held in covered Living Streams, were subjected to control and manipulation of daylength and water temperature. The degree of smoltification under various experimental conditions was assessed by quantifying silvery coloration, downstream migratory activity, salinity tolerance, osmoregulatory capability, and gill Na^+/K^+-ATPase activity in the gills according to the procedures described in the preceding chapter. This accentuates the value of quantifying the smolt transformation in that it allows several types of studies to be conducted with economy of time and effort.

The effect of photoperiodic manipulations on circulating levels of hormones was evaluated by performing radioimmunoassays for thyroid hormones and cortisol. Our major objective in conducting this study was to determine if photoperiod manipulations caused hormone levels to fluctuate prior to changes in morphology, physiology, and behavior.

For the daylength study, water temperature was maintained at a constant $6.0° \pm 1$ °C in each of three tanks from January through mid-July. In January one tank was kept in the dark, a second placed on naturally in-

Table 6.1. Effect of accelerating daylength on smolt transformation and thyroid hormone level in coho salmon

		Accelerated daylength		Normal January daylength	
Before	Cortisol (ng nl^{-1})	1.7 ± 0.4	n = 4	1.4 ± 0.6	n = 5
	T_3 (ng nl^{-1})	1.1 ± 0.3	n = 3	1.3 ± 0.4	n = 5
	T_4 (ng nl^{-1})	9.5 ± 0.9	n = 5	9.1 ± 1.1	n = 11
	Smolt coloration (%)	0	n = 25	0	n = 25
	Migratory behavior (%)	29.7 ± 6.4	n = 25	38.4 ± 9.7	n = 25
	Salinity tolerance (%)	0	n = 10	0	n = 10
	Thyroid histology	Inactive	n = 3	Inactive	n = 3
After 17 days	Cortisol (ng nl^{-1})	11.8 ± 1.8	n = 4	1.8 ± 0.5	n = 3
	T_3 (ng nl^{-1})	10.3 ± 1.0	n = 5	0.9 ± 0.4	n = 5
	T_4 (ng nl^{-1})	67.0 ± 8.1	n = 7	9.6 ± 1.3	n = 5
	Smolt coloration (%)	80.6	n = 15	0	n = 15
	Migratory behavior (%)	93.5 ± 2.5	n = 15	34.3 ± 12.9	n = 15
	Salinity tolerance (%)	73.3	n = 15	6.6	n = 15
	Thyroid histology	Active	n = 3	Inactive	n = 3

Table 6.2. Effect of holding coho salmon at constant daylength or in darkness on the development of natural smolt transformation. This treatment blocked smolt transformation from occurring at the normal time. Levels of thyroid hormones did not increase. Despite being held continously under the same conditions, the fish smolted in mid-July. This was correlated with a surge of thyroid hormones

		Dark		Constant short day		Normal daylength	
January	Cortisol (ng ml^{-1})	No data		No data		1.4	n= 8
	T_3 (ng nl^{-1})	1.2	n= 1	1.4±0.5	n= 2	1.3± 0.4	n= 5
	T_4 (ng nl^{-1})	9.7±1.4	n= 2	9.4±0.9	n= 3	9.1± 1.1	n= 5
	Smolt coloration	0	n=15	0	n=15	0	n= 5
	Migratory behavior (%)	27.1±4.8	n=15	31.1±8.7	n=15	38.4± 9.7	n=25
	Salinity Tolerance (%)	0	n= 5	0	n= 5	0	n=10
May	Cortisol (ng/nl)	No data		No data		40.0	n= 6
	T_3 (ng/nl)	1.1±0.5	n= 2	1.2±0.4	n= 2	10.1± 1.8	n=13
	T_4 (ng/nl)	9.8±1.0	n= 4	11.2±1.1	n= 4	74.1±10.2	n=13
	Smolt coloration (5)	20.0	n=10	10.0	n=10	100.0	n=13
	Migratory behavior (%)	31.7±1.0	n=10	34.3±7.8	n=10	97.6± 1.1	n=13
	Salinity tolerance (%)	40.0	n= 5	60.0	n=10	100.0	n=10
July	Cortisol (ng/nl)	No data		No data			
	T_3 (ng/nl)	6.8±1.3	n= 2	8.1±1.5	n= 3		
	T_4 (ng/nl)	74.1±3.8	n= 4	67.2±8.8	n= 4		
	Smolt coloration (%)	80.0	n= 5	60.0	n= 5		
	Migratory behavior	90.5±3.8	n= 5	89.7±4.0	n= 5		
	Salinity tolerance (%)	100.0		100.0			

creasing daylength (controls) and a third subjected to an artificially advanced and accelerated daylength for 17 days. The degree of smolt transformation and levels of thyroid hormones and cortisol were assessed in all three groups.

After the completion of this experiment in February, the third tank was switched back to natural daylength and some of the fish previously held under natural photoperiod were placed in it. From this point onward the third tank was kept continuously at a constant February daylength. Hence, fish were held under natural daylength, at constant February daylength, or in darkness. This regime was maintained until May when smolt transformation normally occurs and all three groups were tested for degree of smolt transformation and levels of thyroid hormone and cortisol. Fish held in constant February photoperiod and in darkness were tested again in mid-July.

Salmon that were exposed to an accelerated daylength in January displayed elevated levels of thyroid hormones and cortisol and increased silvering, downstream migratory activity, and salinity tolerance compared with fish kept in tanks with natural daylength (Table 6.1). Holding the fish at constant daylength or in dark through May blocked normal smolt transformation; thyroid hormones did not increase at the normal time (Table 6.2). Although still under a constant daylength, or in darkness, the fish smolted in mid-July after a surge of thyroid hormones (Table 6.2).

These data imply that the mechanisms responsible for smolt transformations are entirely internal. The annual photocycle functioning as a Zeitgeber synchronizes or readjusts this endogenous rhythm. Hormones mediate the developmental program and regulate the morphological, physiological, and behavioral transitions, including the olfactory imprinting process.

Postscript

Knowledge of basic fish behavior and physiology is important for managing salmon stocks, since such knowledge enhances possibilities of protecting the world's dwindling populations of salmon. Our studies on olfactory imprinting and homing in salmon illustrate how a basic research can be applied to fisheries management.

Results from our experiments imply that the olfactory identification of the home stream by salmon is learned rapidly and retained until adulthood. Moreover, artificial scents can be used for imprinting in place of natural odors. It is evident, therefore, that the final stages in the migration of some species of salmonids can be manipulated by artificially imprinting smolts to a synthetic chemical (Scholz et al. 1975, Hasler and Scholz 1980).

With respect to Lake Michigan salmonids, these findings suggest several applications:

1. Artificial imprinting in hatcheries reduces the need for smolting ponds, an advantage both in operating cost and survival of young fish. Additionally, water quality and temperatures can be carefully controlled in a hatchery.

2. Artificial imprinting permits manipulation of salmon and trout runs by selecting specific locations for harvest, whereas fish stocked in smolting ponds return only to the river of release. In addition, fish can be stocked at sites within a metropolitan area that does not have adequate facilities for smolting ponds, thereby insuring a lively fishery 18 months later for metropolitan residents. The Michigan Department of Natural Resources, for example, currently uses artificial imprinting techniques to successfully manage their coho and chinook salmon fishery in the Detroit River. By careful planning, such programs can spread out or concentrate sport-fishing pressures as may be desirable.

3. Artificial imprinting can improve the returns and harvest of steelhead trout and other species that normally must be stocked directly in the lake because of unsuitable stream temperatures at time of stocking. Unless they are imprinted artificially, such fish exhibit a poor return to the

stocking site and a high degree of straying, since they are not exposed to a unique odor during the imprinting period (Scholz et al. 1978 a).

Our findings may also be valuable with respect to conservation of stocks of salmon residing in the Atlantic or Pacific Ocean where fishery biologists could use artificial imprinting for attracting adult salmon to suitable areas for spawning or commercial harvest, or diverting them from power dams, warm-water discharges or other potential hazards. For example, in the Columbia River, artificial imprinting could be used to re-establish spawning streams where dams or pollution have eliminated ancestral stocks. At several locations fish ladders have been added as afterthoughts, so it might be possible to attract salmon that have been artificially imprinted in a hatchery to such sites. Once spawning salmon are lured to these streams, perhaps too remote for direct stocking, the offspring would be naturally imprinted to their home-stream system. In addition, fish may be attracted to specific sites along the coast for harvest before they enter the river, hence utilizing them in prime condition at their best commercial value.

An alternative approach would be to utilize chemical imprinting in conjunction with spawning channels. In some areas on the Columbia River artificial channels have been built below a dam in order to provide suitable spawning beds for the fish (e.g., Priest Rapids Dam). Water is pumped from the dam into the channel. Salmon fingerlings were stocked in these spawning channels with the hope that adult fish would return there to spawn and thereby establish a spawning run. The method has not proven very successful because fish do not re-enter the channel at the time of spawning but stray instead into areas which are not suitable for spawning (Allen and Meiken 1973). One possible reason for this may be because the odor of the water coming from the channel is identical to the river water, so that the fish are not able to distinguish the entrance to the spawning channel. It should be possible to scent the channel with a synthetic chemical when the young fish are stocked and later, when the adults return to spawn, to provide a unique marker by re-scenting the channel.

Indeed, much of this promise is being realized. The widespread adaptation of our imprinting techniques in fish-management programs – from Scotland to New Zealand – has been a rewarding outgrowth of basic research on anadromous fish migration.

Our investigations on the mechanism of the imprinting process have other management applications. For example, information about the timing of imprinting or external indicators of imprinting are important for determining the time for stocking fish to insure homing to the stream of release. Also, it might be possible to control hormone levels in hatcheries by manipulating daylength to accelerate smolt transformation, thereby reducing the total cost of raising the fish.

Surely any step toward saving salmon is worthwhile, for they are a gift from nature to man. We neither feed them nor take care of them in any other way. Yet, as young fish, they go out to sea where they grow fat on the ocean's "rangeland" and then deliver themselves, free of charge, to our back door for the catching and eating.

References

Adams BL, Zaugg WS, McLain LR (1973) Temperature effect on parr-smolt transformation in steelhead trout (*Salmo gairdneri*) as measured by gill sodium – potassium stimulated adenosine triphosphate. Comp Biochem Physiol 44A:1333–1339

Adams BL, Zaugg WS, McLain LR (1975) Inhibition of salt water survival and Na^+/K^+-ATPase elevation in steelhead trout (*Salmo gairdneri*) by moderate water temperature. Trans Am Fish Soc 104:766–679

Aho R (1975) Return of Deschutes River summer steelhead to the sports fishery from smolt releases in various locations. 1975 Symposium on Salmon Homing (summary notes). Natl Mar Fish Serv, Seattle, Washington

Ali MA, Hoar WS (1959) Retinal responses of pink salmon associated with its downstream migration. Nature (London) 184:106–107

Allen R, Meiken T (1973) An evaluation of the Preist Rapids chinook spawning channel. Wash Dep Fish Tech Rep No. 11

Allison LN (1951) Delay of spawning in eastern brook trout by means of artificially prolonged light intervals. Prog Fish Cult 13(3):111–116

Arnold GP (1974) Rheotropism in fishes. Biol Rev 49:515–576

Baggerman B (1960a) Factors in the diadromous migrations of fish. Symp Zool Soc London 1:33–58

Baggerman B (1960b) Salinity preference, thyroid activity and the seaward migration of four species of Pacific salmon (*Oncorhynchus*). J Fish Res Board Can 17(3):295–322

Bakshtansky AL, Barybina IA, Nesterov VD (1976a) Changes in the intensity of downstream migration of Atlantic salmon smolts according to abiotic conditions. J Cons Perm Int Explor Mer 4:1–12

Bakshtansky AL, Nesterov VD, Nekludov MN (1976b) Predator's effect on the behaviour of Atlantic salmon smolts in the period of downstream migration. J Cons Perm Int Explor Mer 3:1–19

Bams RA (1976) Survival and propensity for homing as affected by presence or absence of locally adapted paternal genes in two transplanted populations of pink salmon (*Oncorhynchus gorbuscha*). J Fish Res Board Can 33:2716–2725

Banks JH (1969) A review of the literature on the upstream migration of adult salmonids. J Fish Biol 1:85–136

Barnaby JT (1944) Fluctuations in abundance of red salmon, *Oncorhynchus nerka* (Walbaum), of the Karluk River, Alaska. Fish Bull 50(39):237–295

Bentley WW, Raymond HL (1975) Delayed migrations of yearling chinook salmon since completion of Lower Monumental and Little Goose Dams on the Snake River. Trans Am Fish Soc 105:422–424

Bently PJ (1971) Endocrines and osmoregulation. Springer, Berlin Heidelberg New York

Bern HA (1975) Prolactin and osmoregulation. Am Zool 15:937–948

Bern HA (1978) Endocrinological studies on normal and abnormal salmon smoltification. In: Gaillard PJ, Boer HH (eds) Comparative endocrinology. Elsevier/North Holland Biomedical Press, Amsterdam New York, pp 97–100

Black VS (1951) Changes in body chloride, density, and water content of chum (*Oncorhynchus keta*) and coho (*O. kisutch*) when transferred from freshwater to seawater. J Fish Res Board Can 8:164–177

Bodznick D (1975) The relationship of the olfactory EEG evoked by naturally occurring stream waters to the homing behavior of sockeye salmon (*Oncorhynchus nerka* Walbaum). Comp Biochem Physiol 52A:487–495

Boeuf G, Laserre P, Harache Y (1978) Osmotic adaptation of *Oncorhynchus kisutch* Walbaum. II. Plasma osmotic and ionic variations, and gill Na^+-K^+ ATPase activity of yearling coho salmon transferred to sea water. Aquaculture 15:35–52

Bohus B, Gispen W, Wied De D (1973) Effect of lysine vasopressin and ACTH 4–10 on conditioned avoidance behavior of hypophysectomized rats. Neuroendocrinology 11:137–143

Breton B, Billard R (1977) Effects of photoperiod and temperature on plasma gonadotropin and spermatogenesis in the rainbow trout, *Salmo gairdneri* Richardson. Ann Biol Anim Biochem Biophysiol 17:1–10

Brett JR, Groot C (1963) Some aspects of olfactory and visual responses in Pacific salmon. J Fish Res Board Can 20:287–303

Brett JR, MacKinnon D (1954) Some aspects of olfactory perception in migrating adult coho and spring salmon. J Fish Res Board Can 11:310–318

Buckland J (1880) Natural history of British fishes. Unwin, London

Butler D (1973) Structure and function of the adrenal gland in fishes. Am Zool 13:839–881

Carlin B (1955) Tagging of salmon smolts in the River Langan. Inst Freshwater Fish Res, Drottningholm No 36, pp 57–74

Carlin B (1968) Salmon conservation, tagging experiments, and migrations of salmon in Sweden. Lect Ser Atl Salm Assoc, Montreal

Carlson AR, Hale JG (1973) Early maturation of brook trout in the laboratory. Prog Fish Cult 35:150–153

Champney TF, Sahley TL, Sandman CA (1976) Effects of neonatal cerebral ventricular injection of ACTH 4–9 and subsequent adult injections on learning in male and female albino rats. Pharmacol Biochem Behav 5:3–9

Chandrasekhar K, Moskovkin GN, Mitskevich MS (1979) Effect of methythiouracil and triodothyronine on development of the central nervous system in chicken embryos. Gen Comp Endocrinol 37(1):6–14

Chapman DW (1962) Aggressive behavior in juvenile coho salmon as a cause of emigration. J Fish Res Board Can 19:1047–1080

Chidester AM (1924) A critical examination of the evidence for physical and chemical influences on fish migration. J Exp Biol 2:79–118

Chrisp EY, Bjorn TC (1978) Parr-smolt transformation and seaward migration of wild and hatchery steelhead trout in Idaho. Federal aid to Fish and Wildlife Restoration Project, Project F-49-12, Salmon and Steelhead Investigation

Clarke WC, Nagahama Y (1977) The effect of premature transfer to seawater on growth and morphology of the pituitary, thyroid, pancreas and interrenal in juvenile coho salmon (*Oncorhynchus kisutch*). Can J Zool 55:1620–1630

Clarke WC, Farmer SW, Hartwell KM (1977) Effect of teleost pituitary growth hormone on growth of *Tilapia mossambica* and on growth and seawater adaptation of sockeye salmon (*Oncorhynchus nerka*). Gen Comp Endocrinol 33:174–178

Clarke WC, Shelbourne JE, Brett JR (1978) Growth and adaptation to sea water in underyearling sockeye (*Oncorhynchus nerka*) and coho (*O. kisutch*) samon subjected to regimes of constant or changing temperature and day length. Can J Zool 56:2413–2421

Clarke WC, Shelbourne JE, Brett JR (1980) Effect of artifical photoperiod cycles, temperature and salinity on growth and smolting in underyearling coho, chinook, and sockeye salmon. Aquaculture 21:78–85

Collins GB (1976) Effects of dams on Pacific salmon and steelhead trout. Mar Fish Rev 38:39–46

Collins GB, Trefethen DS, Groves AB (1962) Orientation of homing salmon. Am Zool 21:10

Conte FP, Wagner HH (1965) Development of osmotic and ionic regulation in juvenile steelhead trout Salmo gairdneri. Comp Biochem Physiol 14:603–620

Conte FP, Wagner HH, Fessler J, Gnose C (1966) Development of osmotic and ionic regulation in juvenile coho salmon *Oncorhynchus kisutch*. Comp Biochem Physiol 18:1–15

Cooper JC, Hasler AD (1973) An electrophysiological approach to salmon homing. Fish Res Board Can Tech Rept No 415, Pac Biol Stn, Nanaimo, British Columbia

Cooper JC, Hasler AD (1974) Electrophysiological evidence for retention of olfactory cues in homing coho salmon. Science 183:336–338

Cooper JC, Hasler AD (1976) Electrophysiological studies of morpholine-imprinted coho salmon (*Oncorhynchus kisutch*) and rainbow trout (*Salmo gairdneri*). J Fish Res Board Can 33:688–694

Cooper JC, Scholz AT (1976) Homing of artificially imprinted steelhead trout. J Fish Res Board Can 33:826–829

Cooper JC, Scholz AT, Horrall RM, Hasler AD, Madison DM (1976) Experimental confirmation of the olfactory hypothesis with artificially imprinted homing coho salmon (*Oncorhynchus kisutch*). J Fish Res Board Can 33:703–710

Craigie EH (1926) A preliminary experiment on the relation of the olfactory sense to the migration of sockeye salmon (*Oncorhynchus nerka*). Trans R Soc Can 20:215–224

Crim LW, Evans DM (1976) Gonadotropic hormone treatment of rainbow trout (*Salmo gairdneri*): plasma hormone profile following a single injection. J Fish Res Board Can 33:2841–2844

Crim LW, Meyer RK, Donaldson EM (1973) Radioimmunoassay of plasma gonadotropin levels in the spawning pink salmon. Gen Comp Endocrinol 21:69–76

Crim LW, Watte EG, Evans DM (1976) The plasma gonadotropin profile during sexual maturation in a variety of salmon fishes. Gen Comp Endocrinol 27:62–70

Dales S, Hoar WS (1954) Effects of thyroxine and thiourea on the early development of chum salmon *Oncorhynchus keta*. Can J Zool 32:244–254

Davenport JW, Gonzales LM (1973) Neonatal thyroxine stimulation in rats. J Comp Physiol Psychol 2:397–408

Davidson FA, Vaughan E, Hutchinson SJ, Pritchard AL (1943) Factors affecting the upstream migration of pink salmon. Ecology 24:149–168

DeLacey AC (1966) Letter to AD Hasler. In: Hasler AD (ed) Underwater guideposts. Univ Wisconsin Press, Madison, pp 79–80

DeLacey AC, Donaldson LR, Brannon EL (1969) Homing behavior of chinook salmon. Contributions. Univ Washington Coll Fish 300:59–60

de Vlaming VL (1974) Environmental and endocrine control of teleost reproduction. In: Schreck CB (ed) Control of sex in fishes. Extens Div, Va Polytech Inst State Univ, Blacksburg, pp 13–83

DeWied D (1964) Influence of anterior pituitary on avoidance learning and escape behavior. Am J Physiol 207:255–259

DeWied D (1966) Inhibitory effect of ACTH and related peptides on extinction of conditioned avoidance behavior in rats. Proc Soc Exp Biol 122:28–32

DeWied D (1969) Effects of peptide hormones on behavior. In: Lederis K, Cooper K (eds) Frontiers in neuroendocrinology. Oxford Univ Press, London, pp 97–140

DeWied D (1973) The role of posterior pituitary and its peptides on the maintenance of conditioned avoidance behavior. In: Marks BH, Wied De D (eds) Hormones and brain function. Plenum Publishing Corp, New York, pp 87–130

Dickhoff WW, Folmar LC, Gorbman A (1978) Changes in plasma thyroxine during the smoltification of coho salmon, *Oncorhynchus kisutch*. Gen Comp Endocrinol 36:229–232

Dizon AE, Horrall RM, Hasler AD (1973a) Long-term olfactory "memory" in coho salmon, *Oncorhynchus kisutch*. Fish Bull 71:315–317

Dizon AE, Horrall RM, Hasler AD (1973b) Olfactory electroencephalographic responses of homing coho salmon, *Oncorhynchus kisutch*, to water conditioned by conspecifics. Fish Bull 71:893–896

Dodson JJ, Leggett WC (1973) Behavior of adult American shad (*Alosa sapidissima*) homing to the Connecticut River from Long Island Sound. J Fish Res Board Can 30:1187–1860

Dodson JJ, Leggett WC (1974) Role of olfaction and vision in the behavior of American shad (*Alosa sapidissima*) homing to the Connecticut River from Long Island Sound. J Fish Res Board Can 31:1607–1619

Donaldson EM, Yamazaki F, Dye HM, Philleo WW (1972) Preparation of gonadotropin from salmon (*Oncorhynchus tshawytscha*) pituitary glands. Gen Comp Endocrinol 18:469–481

Donaldson LR (1961) Salmon "homing" pond. Res Fish Seattle 1960, Contrib 139:27–28

Donaldson LR (1965) Salmon returns to the "homing" pond. Res Fish Seattle 1964, Contrib 184:24–28

Donaldson LR, Allen GH (1957) Return of silver salmon *Oncorhynchus kisutch* (Walbaum) to point of release. Trans Am Fish Soc 87:13–22

Dornbush RL, Nikolovski O (1976) ACTH 4–10 and short-term memory. Pharmacol Biochem B 5:69–72

Døving KB, Nordeng H, Oakley B (1974) Single unit discrimination of fish odors released by char (*Salmo alpinus* L.) populations. Comp Biochem Physiol 47A:1051–1063

Dozin-van Roye B, De Nayer Ph (1979) Nuclear triiodothyronine receptor in rat brain during maturation. Brain Res 177:551–554

Drucker B (1972) Some life history characteristics of coho salmon of the Karluk River System, Kodiak Island, Alaska. Fish Bull 70:79–99

Eales JG (1963) A comparative of thyroid function in migrant juvenile salmon. Can J Zool 41:811–824

Eales JG (1965) Factors influencing seasonal changes in thyroid activity in juvenile steelhead trout, *Salmo gairdneri*. Can J Zool 43:719–729

Eayrs JT (1964) Effect of neonatal hypothyroidism on maturation and learning in the rat. Anim Behav 12:195–199

Eayrs JT (1966) Thyroid and central nervous development. In: Scientific basis of medicine annual reviews. Univ London Athlone Press, London, pp 317–339

Ebel WJ (1970) Effect of release location on survival of juvenile fall chinook salmon, *Oncorhynchus tshawytscha*. Trans Am Fish Soc 99:672–676

Ebel WJ, Park DL, Johnsen RC (1973) Effects of transportation on survival and homing of Snake River chinook salmon and steelhead trout. Fish Bull 71:549–563

Ellis CH (1957) Homing of chinook salmon transportet downstream after smolt transformation. Prog Fish Cult 19:205–207

Ellis CH (1970) A return to adult coho salmon demonstrating a high degree of selectivity in homing. Wash Dep Fish Manag Rep, Olympia, Washington

Elson PF (1957) The importance of size in the change from parr to smolt in Atlantic salmon. Can Fish Cult 21:1–6

Elson PF (1962) Predator-prey relationships between fish-eating birds and Atlantic salmon. Bull Fish Res Board Can 133:1–87

Emanuel ME, Dodson JJ (1979) Modification of rheotropic behavior of male rainbow trout (*Salmo gairdneri*) by ovarian fluid. J Fish Res Board Can 36:63–68

Essman WB, Mendoza LA, Hainburgh M (1968) Critical periods of maze acquisition development in euthyroid and hypothyroid rodents. Psychol Rep 23:795–800

Ewing RD, Johnson SL (1978) A simplified procedure for analysis of Na^+/K^+ ATPase. Oreg Dep Fish Wildl Inf Rep Ser, Fish 76–3:1–10

Ewing RD, Johnson SL, Pribble JH, Lichatowich JA (1979) Temperature and photoperiod effects on gill (Na+K)-ATPase activity in chinook salmon (*Oncorhynchus tshawytscha*). J Fish Res Board Can 36:1437–1453

Fagerlund UHM, Donaldson EM (1969) The effect of androgens on the distribution and secretion of cortisol in gonadectomized male sockeye salmon (*Oncorhynchus nerka*). Gen Comp Endocrinol 12:438–448

Fagerlund UHM, McBride JR, Smith M, Tomlinson N (1963) Olfactory perception in migrating salmon. III. Stimulants for adult sockeye salmon (*Oncorhynchus nerka*) in home stream waters. J Fish Res Board Can 20:1457–1463

Farmer GJ, Ritter JA (1978) Seawater adaptation and parr-smolt transformation of juvenile Atlantic salmon, *Salmo salar*. J Fish Res Board Can 35:93–100

Fessler J (1974) Determination of factors limiting efficiency of hatchery production and the population characteristics and life history of Deschutes River summer steelhead. Oreg Wildl Comm Fed Aid Prog Rep Fish F-88-R

Fessler J, Wagner HH (1969) Some morphological and biochemical changes in steelhead trout during parr-smolt transformation. J Fish Res Board Can 26:2823–2841

Fletcher CR (1977) Physical chemistry of osmoregulatory physiology: osmotic and ionic regulation in marine teleosts. J Comp Physiol 1243:149–169

Flood JF, Jarvik ME, Bennett EL, Orme AE (1976) Effect of ACTH peptide fragments on memory formation. Pharmacol Biochem Behav 5:41–51

Foerster RE (1929) An investigation of the life history and propagation of the sockeye salmon (*Oncorhynchus nerka*) at Cultus Lake, British Columbia, No. 1. Introduction and the run of 1925. Contrib Can Biol Fish (NS) 5(1):1–35

Foerster RE (1937) The relation of temperature to seaward migration of young sockeye salmon (*Oncorhynchus nerka*). J Biol Board Can 3(5):421–438

Foerster RE (1938) An investigation of the relative efficiencies of natural and artificial propagation of sockeye salmon (*Oncorhynchus nerka*) at Cultus Lake, British Columbia. J Fish Res Board Can 4(3):151–161

Foerster RE (1952) The seaward-migrating sockeye and coho salmon from Lakelse Lake, 1952. Fish Res Board Can, Pac Prog Rep 93:30–32

Foerster RE (1968) The sockeye salmon, *Oncorhynchus nerka*. Bull Fish Res Board Can 162:422 pp

Folmar LC, Dickhoff WW (1979) Plasma thyroxine and gill Na^+-K^+ ATPase changes during seawater acclimation of coho salmon, *Oncorhynchus kisutch*. Comp Biochem Physiol 63A:329–332

Folmar LC, Dickhoff WW (1980) Parr-smolt transformation (smoltification) and sea water adaptation in salmonids: a review of selected literature. Aquaculture 21:1–37

Fontaine M, Leloup J, Olivereau M (1952) La fonction thyroïdienne de jeune saumon, *Salmo salar* (parr et smolt) et son intervention possible dans la migration d'avalaison. Arch Sci Physiol 6:83–104

Fontaine Y (1975) Hormones in fishes. In: Malin DC, Sargent JR (eds) Biochemical and biophysical perspectives in marine biology, vol II. Academic Press, London New York, pp 139–212

Ford D, Crammer EB (1977) Developing nervous systems in relation to thyroid hormones. In: Graves G (ed) Thyroid hormones and brain development. Raven Press, New York

Fried SM, McCleave JD, LaBar GW (1978) Seaward migration of hatchery-reared Atlantic salmon, *Salmo salar*, smolts in the Penobscot River estuary: riverine movements. J Fish Res Board Can 35:76–87

Funk JD, Donaldson EM (1972) Induction of precocious sexual maturity in pink salmon. Can J Zool 50:1414–1419

Funk JD, Donaldson EM, Dye HM (1973) Induction of precocious sexual development in female pink salmon (*Oncorhynchus gorbuscha*). Can J Zool 51:493–500

Gerking S (1959) The restricted movement of fish populations. Biol Rev 34:221–242

Giles MA, Vanstone WE (1976) Changes in ouabain-sensitive adenosine triphosphatase activity in gills of coho salmon (*Oncorhynchus kisutch*) during parr-smolt transformation. J Fish Res Board Can 33:54–62

Glova GL, McInerney JE (1977) Critical swimming speeds of coho salmon (*Oncorhynchus kisutch*) fry to smolt stages in relation to salinity and temperature. J Fish Res Board Can 34:151–154

Godin JG, Dill PA, Drury DE (1974) Effects of thyroid hormones on behaviour of yearling Atlantic salmon (*Salmo salar*). J Fish Res Board Can 31:1787–1790

Gold PE, McGough JL (1977) Hormones and memory. In: Miller LH, Sandman CA, Kastin AJ (eds) Neuropeptide influences on the brain and behavior. Raven Press, New York, pp 127–143

Gold PE, Van Buskirk R (1976) Effects of postrial hormone injections on memory processes. Horm Behav 7:509–517

Gona AG (1973) Effects of thyroxine thyrotropin, prolactin and growth hormone on the maturation of the frog cerebellum. Exp Neurol 38:494–501

Gona AG, Gona OD (1977) Local action of Thyroxine implants on cerebellar maturation in frog tadpoles. Exp Neurol 57(2):581–587

Goodman C, Scholz AT (1980) Homing of coho salmon transplanted in Lake Michigan. Wis Dep Nat Res Tech Rep (submitted)

Gorbman A (1969) Thyroid function and its control in fishes. In: Hoar WS, Randall DJ (eds) Fish physiology, vol II. Academic Press, London New York, pp 241–272

Gorden MS (1959) Osmotic and ionic regulation in Scottish brown trout and sea trout, *Salmo trutta* (L.). J Exp Biol 36:253–260

Groot C (1965) On the orientation of young sockeye salmon (*Oncorhynchus nerka*) during their seaward migration out of lakes. Behaviour (Suppl) 14:198 pp

Groot C, Simpson K, Todd I, Murray PD, Buxton G (1975) Movements of sockeye salmon (*Oncorhynchus nerka*) in the Skeena River estuary as revealed by ultrasonic tracking. J Fish Res Board Can 32:233–242

Groves AB, Collins GB, Trefethen GB (1968) Roles of olfaction and vision in choice of spawning site by homing adult chinook salmon (*Oncorhynchus tshawytscha*). J Fish Res Board Can 25:867–876

Hahn WE (1976) Electroencephalograph of the olfactory bulb in relation to prespawn homing. Experientia 32:1095–1097

Hallock RJ, Elewell RF, Fry DH Jr (1970) Migrations of adult king salmon *Oncorhynchus tshawytscha*, in the San Joaquin delta as demonstrated by the use of sonic tags. Cal Dep Fish Game Fish Bull 151:1–92

Hamburgh M (1968) An analysis of the action of thyroid hormone on development based on in vivo and in vitro studies. Gen Comp Endocrinol 10:198

Hamburgh M (1969) The role of thyroid and growth hormones in Neurogenesis. In: Moscona AA, Monroy A (eds) Current topics in developmental biology. Academic Press, London New York, pp 109–148

Hara TJ (1967a) Effect of steroid hormones on olfactory activity in fish. Comp Biochem Physiol 22:209–225

Hara TJ (1967b) Electrophysiological studies of the olfactory system of the goldfish, *Carrissius auratus* L. III. Effects of sex hormones on the electrical activity of the olfactory bulb. Comp Biochem Physiol 22:209–226

Hara TJ (1970) An electrophysiological basis for olfactory discrimination in homing salmon: a review. J Fish Res Board Can 29:569–586

Hara TJ, Gorbman A (1967) Electrophysiological studies of the olfactory system of the goldfish. Comp Biochem Physiol 21:185–200

Hara TJ, Ueda K, Gorbman A (1965) Influences of thyroxine and sex hormones upon optically evolked potentials in the optic tectum of goldfish. Gen Comp Endocrinol 5:313–319

Hara TJ, Ueda K, Gorbman A (1967) Electroencephalographic studies of homing salmon. Science 149:884–885

Harden-Jones FR (1968) Fish migration. Arnold Press, London, 325 pp

Hartman WL, Raleigh RF (1964) Tributary homing of sockeye salmon at Brooks and Karluk Lakes, Alaska. J Fish Res Board Can 21:485–504

Hartman WL, Strickland CW, Hoopes DT (1962) Survival and behavior of sockeye salmon fry migrating into Brooks Lake, Alaska. Trans Am Fish Soc 91:133–139

Hartman WL, Heard WR, Drucker B (1967) Migratory behavior of sockeye salmon fry and smolts. J Fish Res Board Can 24:2069–2099

Hasler AD (1966) Underwater guideposts – homing of salmon. Univ Wis Press, Madison, 155 pp

Hasler AD, Scholz AT (1978) Olfactory imprinting in coho salmon. In: Schmidt-Koenig K, Keeton WT (eds) Animal migration, navigation and homing. Springer, Berlin Heidelberg New York, pp 356–369

Hasler AD, Scholz AT (1980) Artificial imprinting: a procedure for conserving salmon stocks. In: Bardach J, Magnuson JJ, May RC (eds) Fish behavior and its use in capture and culture of fishes. Int Cent Liv Aquat Res Manag, Manila, pp 179–199

Hasler AD, Wisby WJ (1951) Discrimination of stream odors by fishes and relation to parent stream behavior. Am Nat 85:223–238

Hasler AD, Scholz AT, Horrall RM (1978) Olfactory imprinting and homing in salmon. Am Sci 66:347–355

Hayes FR (1953) Artificial freshets and other factors controlling the ascent and population of salmon in the Le Have River, Nova Scotia. Bull Fish Res Board Can 99:47 p

Hazard TR, Eddy PE (1951) Modification of the sexual cycle in brook trout, *Salvelinus fontinalis*, by control of light. Trans Am Fish Soc 80:158–162

Heard WR, Crone RA (1976) Raising coho salmon from fry to smolts in estuarine pens, and returns of adults from two smolt releases. Prog Fish Cult 38:171–174

Henderson NE (1963) Influence of light and temperature on the reproductive cycle of the eastern brook trout, *Salvelinus fontinalis* (Mitchill). J Fish Res Board Can 20:859–897

Hess EH (1973) Imprinting. Early experience and the developmental psychobiology of attachment. Van Nostrand Reinhold, New York, 472 pp

Higgs DA, Donaldson EM, Dye HM, McBride JR (1976) Influence of bovine growth hormone and L-thyroxine on growth, muscle composition and histological structure of the gonads, thyroid, pancreas and pituitary of coho salmon (*Oncorhynchus kisutch*). J Fish Res Board Can 33:1585–1603

Hirsch PJ (1977) Conditioning of heart rate of coho salmon (*Oncorhynchus kisutch*) to odors. Ph D Thes, Univ Wis, Madison

Hiyama Y, Taniuchi T, Suyama K, Ishoka K, Sato R, Kajihara T, Maiwa R (1966) A preliminary experiment on the return of tagged chum salmon to the Otsuchi River, Japan. Jpn Soc Sci Fish 33:18–19

Hoar WS (1939) The thyroid gland of the Atlantic salmon. J Morphol 65:257–295

Hoar WS (1951) The behaviour of chum, pink and coho salmon in relation to their seaward migration. J Fish Res Board Can 8:241–263

Hoar WS (1953) Control and timing of fish migration. Biol Rev 28:437–452

Hoar WS (1954) The behaviour of juvenile Pacific salmon with particular reference to the sockeye (*Oncorhynchus nerka*). J Fish Res Board Can 11:69–97

Hoar WS (1958) The evolution of migratory behaviour among juvenile salmon of the genus *Oncorhynchus*. J Fish Res Board Can 15:391–428

Hoar WS (1965) The endocrine system as a chemical link between the organism and its environment. Trans R Soc Can 3:175–200

Hoar WS (1976) Smolt transformation: Evolution, behavior and physiology. J Fish Res Board Can 33:1234–1252

Hoar WS, Bell GM (1950) The thyroid gland in relation to the seward migration of Pacific salmon. Can J Res 28:126–136

Hoar WS, Kennleyside MHA, Goodall RG (1955) The effects of thyroxine and gonadal steroids on the activity of salmon and goldfish. Can J Zool 33:428–439

Hoar WS, Kennleyside MHA, Goodall RG (1957) Reactions of juvenile Pacific salmon to light. J Fish Res Board Can 14:815–830

Hoover EE (1937) Experimental modification of the sexual cycle in trout by control of light. Science 86:425–426

Hoover EE, Hubbard HE (1937) Modification of the sexual cycle in trout by control of light. Copeia 4:206–210

Horrall RM (1981) Behavioral stock-isolating mechanisms in Great Lakes fishes with special reference to homing and site imprinting. Can J Fish Aquat Sci 38:1481–1496

Houston AH (1959) Osmoregulatory adaptations of steelhead (*Salmo gairdneri* Richardson) to sea water. Can J Zool 37:729–748

Houston AH (1960) Variations in the plasma-level of chloride in hatchery-reared Atlantic salmon during parr-smolt transformation and following transfer into sea water. Nature (London) 185:632–633

Houston AH (1961) Influence of size upon the adaptation of steelhead trout (*Salmo gairdneri*) and chum salmon (*Oncorhynchus keta*) to sea water. J Fish Res Board Can 18:401–415

Houston AH (1964) On passive features in the osmoregulatory adaptation of anadromous salmonids to sea water. J Fish Res Board Can 21:1535–1542

Houston AH, Threadgold LT (1963) Body fluid regulation in smolting Atlantic salmon. J Fish Res Board Can 20:1355–1369

Hunter JG (1959) Survival and production of pink and chum salmon in a coastal stream. J Fish Res Board Can 16:835–886

Huntsman AG (1942) Return of a marked salmon from a distant place. Science 95:381–382

Huntsman AG (1952) Wandering versus homing in salmon. Salm Trout Mag 136:185–191

Huntsman AG, Hoar WS (1939) Resistance of Atlantic salmon to sea water. J Fish Res Board Can 4:409–411

Idler DR, Ronald P, Schmidt PJ (1959) Biochemical studies on the sockeye salmon during spawning migrations. VII. Steroid hormones in plasma. Can J Biochem 37:1227–1238

Idler DR, McBride JR, Jones REE, Tomlinson N (1961) Olfactory perception in migrating salmon. II. Studies on a laboratory bio-assay for homestream water and mammalian repellent. Can J Biochem Physiol 39:1575–1584

Isaksson A (1976) The improvement of returns of one-year smolts at the Kollafjördur Fish Farm 1971–73. J Agric Res Icel 8:19–26

Ishiguro K, Suzuki Y, Sato T (1980) Effect of neonatal hypothyroidism on maturation of nuclear triiodothhornine (T_3) receptors in developing rat brain ACTA. Endocrinology 95:459–499

Jahn LA (1966) Open-water movements of cutthroat (*S. clarki*) in Yellowstone Lake after displacement from spawning stream. J Fish Res Board Can 23:10–17

Jahn LA (1969) Movements and homing of cutthroat from open water areas of Yellowstone Lake. J Fish Res Board Can 26:1243–1261

Jahn LA (1976) Responses to odors by fingerling cutthroat trout from Yellowstone Lake. Prog Fish Cult 4:207–210

Jensen A, Duncan R (1971) Homing in transplanted coho salmon. Prog Fish Cult 33:216–218

Johnsen PB (1978) Behavioral mechanisms of upstream migration and home stream selection in coho salmon (*Oncorhynchus kisutch*). Ph D Thes, Univ Wis, Madison, 120 pp

Johnsen PB, Hasler AD (1980) The use of chemical cues in the upstream migration of coho salmon, *Oncorhynchus kisutch*, Walbaum. J Fish Biol 17:67–73

Johnson DW (1973) Endocrine control of hydromineral balance in teleosts. Am Zool 13:799–818

Johnson SL, Ewing RD, Lichatowich JA (1977) Characterization of gill Na^+/K^+ activated adenosine triphosphatase from chinook salmon gill tissue. J Exp Zool 199:345–354

Johnson WE, Groot C (1963) Observations on the migration of young sockeye salmon (*Oncorhynchus nerka*) through a large, complex lake system. J Fish Res Board Can 20:919–938

Johnston CE, Eales JG (1967) Purines in the integument of the Atlantic salmon (*Salmo salar*) during parr-smolt transformation. J Fish Res Board Can 24:955–964

Johnston CE, Eales JG (1968) Influence of temperature and photoperiod on guanine and hypoxanthine levels in skin and scales of Atlantic salmon (*Salmo salar*) during parr-smolt transformation. J Fish Res Board Can 25:1901–1909

Johnston CE, Eales JG (1970) Influence of body size on silvering of Atlantic salmon (*Salmo salar*) at parr-smolt transformation. J Fish Res Board Can 27:983–987

Kaji S, Satou M, Kudo Y, Ueda K, Gorbman A (1975) Spectral analysis of olfactory responses of adult spawning chum salmon to stream waters. Comp Biochem Physiol 51A:711–716

Kalleberg H (1958) Observations in a stream tank of territoriality and competition in juvenile salmon and trout (*Salmo salar* L. and *S. trutta* L.). Rep Inst Freshwater Res Drottningholm 39:55–98

Keenleyside MH, Hoar WS (1954) Effects of temperature on the responses of young salmon to water currents. Behaviour 7:77–87

Kerstetter CE, Keeler M (1976) Smolting in steelhead trout *Salmo gairdneri:* A comparative study of populations in two hatcheries and the Trinity River, Northern California, using gill Na^+/K^+ ATPase assays. Humbolt State Univ Publ SG-9:26 pp

King G, Swanson BL (1973) Marking of rainbow trout in lake Superior. Wisconsin Dept Nat Res, Int Rep, 14 pp

Kleerekoper H (1969) Olfaction in fishes. Indiana Univ Press, Bloomington, 222 pp

Koch HJ, Evans JC, Bergstrom E (1959) Sodium regulation in the blood of parr and smolt stages of Atlantic salmon. Nature (London) 184:283

Komourdjian MP (1976) Photoperiod and the anterior pituitary hormones in growth and development of the Atlantic salmon, *Salmo salar* L., with particular emphasis on growth hormone and prolactin. Ph D Thes, Univ Ottawa, Ontario, 148 pp

Komourdjian MP, Saunders RL, Fenwick JC (1976a) The effect of percine somatotropin on growth and survival in sea water of Atlantic salmon (*Salmo salar*) parr. Can J Zool 54:531–535

Komourdjian MP, Saunders RL, Fenwick JC (1976b) Evidence for the role of growth hormone as a part of a light-pituitary axis in growth and smoltification of Atlantic salmon (*Salmo salar*). Can J Zool 54:544–551

Krokhin EM (1975) Transport of neutrients by Salmon migrating from the sea into lakes. In: Hasler AD (ed) Coupling of land water ecosystems. Springer, Berlin Heidelberg New York, pp 153–156

Kubo T (1953) On the blood of salmonid fishes of Japan during migration. I. Freezing point of blood. Bull Fac Fish, Hokkaido Univ 4:138–148

Kubo T (1955) Changes in some characteristics of blood of smolts of *Oncorhynchus masu* during seaward migration. Bull Fac Fish, Hokkaido Univ 6:201–207

LaBar GW (1971) Movement and homing of cutthroat trout (*S. clarki*) in Clear and Bridge Creeks of Yellowstone National Park. Trans Am Fish Soc 100:41–49

LaBar GW, McLeave JD, Fried SM (1978) Seaward migration of hatchery-reared Atlantic salmon (*Salmo salar*) smolts in the Penobscot River estuary, Maine: Open-water movements. J Cons Perm Int Explor Mer 38:257–270

Lander JM (1979) Granule cell migration in developing rat cerebellum. Influence of neonatal hypo- and hyper-thyroidism. Dev Biol 70(1):105–115

Landgrebe FW (1941) The role of the pituitary and the thyroid in the development of teleosts. J Exp Biol 18:162–169

LaRoche G, Leblond CP (1954) Destruction of the thyroid gland of Atlantic salmon, *Salmo salar* L., by means of radioiodine, Proc Soc Exp Biol Med 87:273–276

LaRoche G, Woodall AN, Johnson CL, Halver JE (1966) Thyroid function in the rainbow trout: Effects of thyroidectomy on young fish. Gen Comp Endocrinol 6:249–266

Larson RW, Ward JM (1954) Management of steelhead trout in the State of Washington. Trans Am Fish Soc 84:261–274

Lasserre P, Boeuf G, Harache Y (1978) Osmotic adaptation of *Oncorhynchus kisutch* Walbaum. I. Seasonal variations of gill Na^+-K^+ ATPase activity in coho salmon, 0^+-age and yearling, reared in fresh water. Aquaculture 14:365–382

Leatherland JF, Lin L (1975) Activity of the pituitary gland in embryo and larval stages of coho salmon, *Oncorhynchus kisutch*. Can J Zool 53:297–310

Leatherland JF, McKeown BA (1973) Effect of ambient salinity on prolactin and growth hormone secretion and on hydro-mineral regulation in kokanee salmon smolts (*Oncorhynchus nerka*). J Comp Physiol 89:215–226

Leatherland JF, McKeown BA (1974) Effect of ambient salinity on prolactin and growth hormone secretion and on hydro-mineral regulation in kokanee salmon smolts (*Oncorhynchus nerka*). J Comp Physiol 89:215–226

Leggett WC (1967) A study of the rate and pattern of shad migration in the Connecticut River utilizing sonic tracking apparatus. US Bur Commerc Fish I. Progr Rep, Project AFC-1-1:13 p

Legrand J (1965) Influence de l'hypothyroïdisme sur la maturation du cortex cérébelleux. C R Acad Sci (Paris) 261:544–547

Legrand J (1967a) Analyse de l'action morphogénétique des hormones thyroïdennes sur le cervelet du jeune rat. Arch Anat Microsc Morphol Exp 56:205–244

Legrand J (1967b) Variations en fonction de l'age de la reponse du cervelot a l'action morphogénétique de la thyroïde chez le rat. Arch Anat Microsc Morphol Exp 56:291–307

Legrand J (1979) Morphogenetic actions of thyroid hormones. Trends Neurosci 2(9):234–236

Legrand J, Kriegel A, Jost A (1961) Deficience thyroidienne et maturation du cervelet chez le rat blanc. Arch Anat Microsc Morphol Exp 50:407–419

Levine S (1968) Hormones and conditioning. In: McGaugh JL, Whalen RE (eds) Psychobiology, the biological basis of behavior. Freeman, San Francisco, pp 85–101

Liley NR (1982) Chemical communication in fish. Can J Fish Aquat Sci 39:22–35

Lindsay CC, Northcote TG, Hartman GF (1959) Homing of rainbow trout to inlet and outlet spawning streams at Loon Lake, British Columbia. J Fish Res Board Can 16:695–719

Lorz HW, McPherson BP (1976) Effects of copper or zinc in freshwater on adaptation of sea water and ATPase activity, and the effects of copper on migratory disposition in coho salmon (*Oncorhynchus kisutch*). J Fish Res Board Can 33:2023–2030

Lorz HW, Northcote TG (1965) Factors affecting stream location and timing and intensity of entry by spawning kokanee (*O. nerka*) into an inlet of Nicola Lake, British Columbia. J Fish Res Board Can 22:665–687

Lovern JA (1934) Fat metabolism in fishes. V. The fat of the salmon in its young freshwater stages. Biochem J 28:1961–1963

MacKinnon CN, Donaldson EM (1976) Environmentally induced precocious sexual development in the male pink salmon (*Oncorhynchus gorbuscha*). J Fish Res Board Can 33:2602–2605

MacKinnon CN, Donaldson EM (1978) Comparison of the effect of salmon gonadotropin administered by pellet implantation or injection on sexual development of juvenile male pink salmon (*Oncorhynchus gorbuscha*). Can J Zool 56:86–89

Madison DM, Horrall RM, Stasko AB, Hasler AD (1972) Migratory movements of adult sockeye salmon (*Oncorhynchus nerka*) in coastal British Columbia as revealed by ultrasonic tracking. J Fish Res Board Can 29:1025–1033

Madison DM, Scholz AT, Cooper JC, Hasler AD (1973) I. Olfactory hypotheses and salmon migration: a synopsis of recent findings. Fish Res Board Can Tech Rep No 414:37 pp

Maetz J (1971) Fish gills: mechanisms of salt transfer in fresh water and sea water. Trans R Soc London Ser B 262:209–249

Mahnken C, Joyner T (1975) Salmon for New England fisheries. III. Developing a coastal fishery for Pacific salmon. Mar Fish Rev (Pap 1010) 35:9–13

Manteifel BP, Girsa II, Parlor DS (1978) On rhythms of fish behaviour. In: Thorpe JE (ed) Rhythmic activity of fishes. Academic Press, London New York, pp 68–75

Martin JT (1978) Imprinting behavior: Pituitary-adrenocortical modulation of the approach response. Science 200:565–566

Mayer N (1970) Cortisol enhancement of sodium-potassium ATPase activity in teleosts adapted to salt water. Nature (London) 214:1118–1120

Mayr E (1974) Behavior programs and evolutionary strategies. Am Sci 62:650–659

McBride JR, Fagerlund UHM, Smith M, Tomlinson N (1963) Resumption of feeding by and survival of adult sockeye salmon (*Oncorhynchus nerka*) following advanced gonad development. J Fish Res Board Can 20:95–100

McBride JR, Fagerlund UHM, Smith M, Tomlinson N (1964) Olfactory perception in juvenile salmon. II. Conditioned response of juvenile sockeye salmon (*Oncorhynchus nerka*) to lake water. Can J Zool 42:45–48

McCart P (1967) Behavior and ecology of sockeye salmon fry in the Babine River. J Fish Res Board Can 24:375–428

McCleave JD (1967) Homing and orientation of cutthroat trout (*S. clarki*) in Yellowstone Lake with special reference to olfaction and vision. J Fish Res Board Can 24:2011–2044

McCleave JD (1978) Rhythmic aspects of estuarine migration of hatchery reared Atlantic salmon, *Salmo salar* smolts. J Fish Biol 12:559–571

McCleave JD, Horrall RM (1970) Ultrasonic tracking of homing cutthroat trout (*Salmo clarki*) in Yellowstone Lake. J Fish Res Board Can 27:715–730

McCleave JD, LaBar G (1972) Further ultrasonic tracking and tagging studies of homing cutthroat trout (*Salmo clarki*) in Yellowstone Lake. Trans Am Fish Soc 101:44–54

McCleave JD, Stred KA (1975) Effect of dummy telemetry transmitters on stamina of Atlantic salmon (*Salmo salar*) smolts. J Fish Res Board Can 32:559–563

McInerney JE (1964) Salinity preference: an orientation mechanism in salmon migration. J Fish Res Board Can 21:995–1018

Mighell J (1975) Some observations of imprinting of juvenile salmon in fresh and saltwater. 1975 Symposium on Salmon Homing (summary notes). Natl Mar Fish Ser, Seattle, Washington

Mills D (1971) Salmon and trout: A resource, its ecology, conservation and management. St. Martins Press, London, 235 pp

Mills D, Shackley PE (1971) Salmon smolt transportation experiments on the Canon River system, Ross-Shire. Freshwater Salm Fish Res 50:58–60

Mondon PM, Kaltenbach JL (1979) Thyroxine concentration in blood serum and pericardial fluid of metamorphosing tadpoles and of adult frogs. Gen Comp Endocrinol 39:343–349

Naidoo S, Valcana T, Timiras PS (1978) Thyroid hormone receptors in developing rat brain. Am Zool 18:545–552

Neave F (1966) Salmon of the North Pacific Ocean: (6) Chum salmon in British Columbia. Bull Int Pac Salm Fish Comm 18:81–86

Nordeng H (1971) Is the local orientation of anadromous fish determined by pheromones? Nature (London) 233:411–413

Nordeng H (1977) A pheromone hypothesis for homeward migration in anadromous salmonids. Oikos 28:155–159

Northcote TG (1958) Effect of photoperiodism on the response of juvenile trout to water currents. Nature (London) 181:1283–1284

Northcote TG (1962) Migratory behavior of juvenile rainbow trout in outlet and inlet streams of Loon Lake, British Columbia. J Fish Res Board Can 19:201–270

Ogawa M (1974) The effects of bovine prolactin, sea water and environmental calcium on water influx in isolated gills of the euryhaline teleosts, *Anguilla japonica* and *Salmo gairdneri*. Comp Biochem Physiol 49A:545–553

Olivereau M (1954) Hypophyse et glande thyroïde chez les poissons. Etude histophysiologique de quelques corrélations endocriniennes, en particulier chez *Salmo salar* L. Ann Inst Oceanogr N S 29:95–296

Oppenheimer JH, Schwartz HL, Surks MI (1974) Tissue differences in the concentration of triiodothyronine nuclear binding sites in the rat: liver, kidney, pituitary, heart, brain, spleen and testis. Endocrinology 95:897–903

Osborn RH, Simpson TH, Youngson AF (1978) Seasonal and diurnal rhythms of thyroidal status in the rainbow trout, *Salmo gairdneri* Richardson. J Fish Biol 12:531–540

Oshima K, Gorbman A (1966a) Olfactory responses in the forebrain of goldfish and their modification by thyroxine treatment. Gen Comp Endocrinol 7:398–409

Oshima K, Gorbman (1966b) Influence of thyroxine and steroid hormones on spontaneous and evoked unitary activity in the olfactory bulb of goldfish. Gen Comp Endocrinol 7:482–491

Oshima K, Gorbman A (1968) Modification by sex hormones of the spontaneous and evoked bulbar electrical activity in goldfish. J Endocrinol 40:409–420

Oshima K, Gorbman A, Shimada H (1969a) Memory-blocking agents: Effects on olfactory discrimination in homing salmon. Science 165:86–88

Oshima K, Hahn WE, Gorbman A (1969b) Olfactory discrimination of natural waters by salmon. J Fish Res Board Can 26:2111–2121

Oshima K, Hahn WE, Gorbman A (1969c) Electroencephalographic olfactory responses in adult salmon to waters traversed in the homing migration. J Fish Res Board Can 26:2123–2133

Oshima K, Johnson CL, Gorbman A (1972) Relations between prolonged hypothyroidism and electroneurophysiological events in trout, *Salmo gairdnerii*. Effects of replacement dosages of thyroxine. Gen Comp Endocrinol (Suppl) 3:529–541

Oshima K, Hoshai G, Tarby T, Gorbman A (1973) Electroencephalographic studies on homing mechanisms in chum salmon. In: Sato R (ed) An experimental study of homing behavior of chum salmon. Final Report to US-Japan Cooperative Science Program

Osterdahl L (1969) The smolt run of a small Swedish river. In: Northcote TG (ed) Salmon and trout in streams. MacMillan HR Lectures in Fisheries. Univ British Columbia Press, Vancouver, pp 205–215

Otto RG, McInerney JE (1970) Development of salinity preference in pre-smolt coho salmon, *Oncorhynchus kisutch*. J Fish Res Board Can 27:793–800

Park D (1975) Homing of salmon transported from Little Goose Dam to Bonneville Dam. 1975 Symp Salm Hom (summary notes). Natl Mar Fish Serv, Seattle, Washington

Parry G (1960) The development of salinity tolerance in the salmon, *Salmo salar* (L.) and some related species. J Exp Biol 37:425–434

Patel AJ, Lewis PD, Balaz R, Bailey P, Lai M (1979) Effects of T_4 on postnatal cell acquisition in the rat brain. Brain Res 172(1):57–72

Peck JW (1970) Straying and reproduction of coho salmon, *Oncorhynchus kisutch*, planted in a Lake Superior tributary. Trans Am Fish Soc 99:591–595

Peters M (1971) Sensory mechanisms of homing in salmonids: A comment. Behavior 39:18–19

Poston HA (1978) Neuroendocrine mediation of photoperiod and other environmental influences on physiological responses in salmonids: A review. U S Fish Wildl Serv Tech Rep No 96:14 pp

Potts WT (1970) Osmotic and ionic regulation. Annu Rev Physiol 30:73–104

Pritchard AL (1938) Transplantation of pink salmon (*O. gorbuscha*) into Masset Inlet, British Columbia, in the barren years. J Fish Res Board Can 6:392–398

Pritchard AL (1943) Results of pink salmon marking at Morrison Creek, British Columbia. B C Fish Res Board Can Prog Rep No 57:8–11

Rabie A, Favre C, Clavel MC, Legrand J (1979) Sequential effects of thyroxine on the developing cerebellum of rats made hypothyroid by propylthiouricil. Brain Res 161:469–479

Ramsey DA (1961) Olfactory cues in migrating salmon. Science 133:56–57

Reimers PE (1979) Success in a hatchery program with full chinook Salmon by simulating the natural life history of the stock. Prog Fish Cult 41:192–195

Reingold M (1975a) Experimental rearing of anadromous fish, Lemhi River drainage, Idaho (Hayden Creek Research Station) July 1, 1970-June 30, 1974. Idaho Fish Game Dep, 18 pp

Reingold M (1975b) Evaluation of transplanting Snake River steelhead trout to the Pahsimeroi River, 1974. Annu Progr Rep, Project IPC-26. Idaho Fish Game Dep, 20 pp

Rich WH, Holmes HB (1929) Experiments in marking young chinook salmon of the Columbia River. Fish Bull 44:215–264

Ricker WE (1972) Hereditary and environmental factors affecting certain salmonid populations. In: Simon RC, Larkin PA (eds) The stock concept in pacific salmon. MacMillan HR Lectures in Fisheries. Univ British Columbia, Vancouver, pp 27–160

Robertson OH (1948) The occurrence of increased activity of the thyroid gland in rainbow trout at the time of transformation from parr to silvery smolt. Physiol Zool 21:282–294

Robertson OH (1949) Production of the silvery smolt stage in rainbow trout by injection of mammalian thyroid extract and thyrotropic hormone. J Exp Zool 110:337–355

Robertson OH, Rinfret AP (1957) Maturation of the infantile testis in rainbow trout produced by salmon pituitary gonadotropins administered in cholesterol pellets. Endocrinology 60:559–562

Rounsefell GA, Kelez GB (1938) The salmon and salmon fisheries of Swiftsure Bank, Puget Sound, and the Fraser River. Bull U S Bur Fish 49:693–823

Sakano E (1960) Results from marking experiments on young chum salmon, 1951–1959. Sci Rept Hakkaido Salm Hatchery No 15:17–38

Sandoval WA (1980) Odor detection by coho salmon (*O. kisutch*): a laboratory bioassay and genetic basis. M S Thes, Oreg State Univ, 43 pp

Sano S (1966) Salmon of the North Pacific Ocean: (3) Chum salmon in the Far East. Bull Int North Pac Salm Fish Comm 18:41–58

Sato R, Hiyama Y, Kajihuma T (1966) The role of odor in return of chum salmon, *Oncorhynchus keta*, to its parent stream. Proc 11 Pac Sci Congr, Tokyo 1966, vol 7:20

Satou M, Ueda K (1975) Spectral analysis of olfactory responses in rainbow trout, *Salmo gairdneri*. Comp Biochem Physiol 52A:359–365

Saunders RL (1965) Adjustment of buoyancy in young Atlantic salmon and brook trout by changes in swim-bladder volume. J Fish Res Board Can 22:335–352

Saunders RL, Henderson EB (1970) Influence of photoperiod on smolt development and growth of Atlantic salmon (*Salmo salar*). J Fish Res Board Can 27:1295–1311

Schalock RL, Brown WJ, Smith RL (1977) Neonatal hypothyroidism, behavioral, thyroid hormonal and neuroanatomical effects. Physiol Behav 19(4):489–491

Schapiro S (1968) Some physiological, biochemical, and behavioral consequences of neonatal hormone administration: Cortisol and thyroxine. Gen Comp Endocrinol 10:214–228

Scheer BT (1939) Homing instinct in salmon. Q Rev Biol 14(4):408–420

Schmidt PJ, Mitchell BS, Smith M, Tsuyuki H (1965) Pituitary hormone of the Pacific salmon. I. Response of gonads in immature trout (*Salmo gairdneri*) to extracts of pituitary glands from adult Pacific salmon (*Oncorhynchus*). Gen Comp Endocrinol 5:197–206

Scholz AT (1980) Hormonal regulation of smolt transformation and olfactory imprinting in coho salmon. Ph D Thes, Univ Wis, 363 pp

Scholz AT, Madison DM, Stasko AB, Horrall RM, Hasler AD (1972) Orientation of salmon in response to water currents in or near the homestream. Am Zool 12:54

Scholz AT, Cooper JC, Madison DM, Horrall RM, Hasler AD, Dizon AE, Poff RJ (1973) Olfactory imprinting in coho salmon: Behavioral and electrophysiological evidence. Proc Conf Great Lakes Res 16:143–153

Scholz AT, Horrall RM, Cooper JC, Hasler AD, Madison DM, Poff RJ, Daly R (1975) Artificial imprinting of salmon and trout in Lake Michigan. Wis Dep Nat Res Fish Manag Rep 80:46 pp

Scholz AT, Horrall RM, Cooper JC, Hasler AD (1976) Imprinting to chemical cues: The basis for homestream selection in salmon. Science 196:1247–1249

Scholz AT, Gosse CK, Cooper JC, Horrall RM, Hasler AD, Daly RI, Poff RJ (1978 a) Homing of rainbow trout transplanted in Lake Michigan: A comparison of three procedures used for imprinting and stocking. Trans Am Fish Soc 107:439–443

Scholz AT, Horrall RM, Cooper JC, Hasler AD (1978 b) Homing of morpholine-imprinted brown trout (*Salmo trutta*). Fish Bull 76:293–295

Schwartz HL, Oppenheimer JH (1978) Ontogenesis of 3, 5, 3′-Triiodothyronine receptors in neonatal rat brain. Endocrinology 103:943–948

Selset R, Døving KB (1980) Behavior of mature anadromous char (*Salma alpinus* L.) towards odorants produced by smolts of their own population. Acta Physiol Scand 108:113–122

Shapovalov L, Taft AC (1954) The life histories of the steelhead rainbow trout (*Salmo gairdneri*) and silver salmon (*Oncorhynchus kisutch*) with special reference to Waddell Creek, California, and recommendations regarding their management. Cal Fish Game, Fish Bull No 98:375 pp

Shearer L (1959) Sea trout transportation experiments. Rep Challenger Soc 3:24–25

Shirahata S, Tanaka M (1969) Homing behavior of land-locked sockeye salmon in Lake Chuzenji. Annu Meet Jpn Soc Sci Fish

Siegel S (1965) Non parametic statistics for the behavioral science. McGraw-Hill, New York, pp 217

Slatick E, Park DL, Ebel WJ (1975) Further studies regarding effects of transportation on survival and homing of Snake River chinook salmon and steelhead trout. Fish Bull 73:925–931

Solomon DJ (1973) Evidence for pheromone-influenced homing by migrating Atlantic salmon, *Salmo salar* L. Nature (London) 244:231–232

Solomon DJ (1975) Observations on some factors influencing the migration of smolts of salmon (*Salmo salar* L.) and migratory trout (*S. trutta* L.) in a chalkstream. J Cons Perm Int Explor Mer 11:1–10

Solomon DJ (1978) Migration of smolts of Atlantic salmon (*Salmo salar* L.) and sea trout (*Salmo trutta* L.) in a chalkstream. Environ Biol Fish 3:255–261

Stasko AB (1971) Review of field studies on fish orientation. Ann N Y Acad Sci 188:12–29

Stasko AB (1975) Progress of migrating Atlantic salmon (*Salmo salar*) along an estuary, observed by ultrasonic tracking. J Fish Biol 7:329–338

Stasko AB, Sutterlin AM, Rommel SA Jr, Elson PF (1973) Migration orientation of Atlantic salmon (*Salmo salar* L.). Int Atl. Salm Symp Spec Ser 4:119–137

Stein RA, Reimers PE, Hall JD (1972) Social interaction between juvenile coho (*Oncorhynchus kisutch*) and fall chinook salmon (*O. tshawytscha*) in Sixes River, Oregon. J Fish Res Board Can 29:1737–1748

Stratton LO, Gibson CA, Kolar KG, Kastin AJ (1976) Neonatal treatment with TRH affects development, learning and emotionality in the rat. Biochem Behav 5:65–67

Stuart TA (1957) The migration and homing behavior of brown trout (*Salmo trutta*). Freshwater Salm Fish Res 18:1–18

Stuart TA (1958) Marking and regeneration of fins. Dep Agric Fish, Scotland. Freshwater Salm Fish Res 22:1–14

Stuart TA (1959) Tenth annual report of the supervisory committee for brown trout research, 1957–1958. Freshwater Salm Fish Res 23:6–7

Sutterlin AM, Gray R (1973) Chemical basis for homing in Atlantic salmon (*Salmo salar*). J Fish Res Board Can 30:985–989

Sutterlin AM, Saunders RL, Hendersen EB, Harmon PR (1982) The homing of Atlantic salmon (*Salmo saler*) to a marine site. Can Tech Rep Fish Aquat Sci 1058:6 pp

Swift DR (1959) Seasonal variation in the activity of the thyroid gland of yearling brown trout (*Salmo trutta Lin.*) J Exp Biol 36:120–125

Tarrant RM (1966) Thresholds of perception of eugenol in juvenile salmon. Trans Am Fish Soc 95:112–115

Teichmann H (1959) Über die Leistung des Geruchssinnes beim Aal (*Anguilla anguilla*). Vergl Physiol 42:206–254

Thommes RC, Vieth RL, Levasseur (1977) The effects of hypophysectomy by means of surgical decapitation on thyroid function in the developing chick embryo. Gen Comp Endocrinol 31:29–36

Thorpe JE, Morgan RIG (1978) Periodicity in Atlantic salmon *Salmo salar* L. smolt migration. J Fish Biol 12:541–548

Tomlinson N, McBride JR, Geiger SE (1967) Sodium, potassium and water content in flesh of sockeye salmon in relation to sexual development and starvation. J Fish Res Board Can 24:243–248

Trevanius GR (1822) Biologie oder Philosophie der lebenden Natur für Naturforscher und Ärzte , vol VI. Rower, Göttingen

Tytler P, Thorpe JE, Shearer WM (1978) Ultrasonic tracking of the movements of Atlantic salmon, *Salmo salar*, in the estuaries of two Scottish rivers. J Fish Biol 12:575–586

Ueda K, Hara TJ, Gorbman A (1967) Electroencephalographic studies on olfactory discrimination in adult spawning salmon. Comp Biochem Physiol 21:133–143

Ueda K, Hara TJ, Satou M, Kaji S (1971) Electrophysiological studies on olfactory discrimination of natural water by sockeye salmon. J Fac Sci Univ Tokyo Sec IV 12:167–182

Ueda K, Satou M, Kaji S, Kudo Y, Gorbman A (1973) Electrophysiological studies on olfactory recognition of home water in spawning chum salmon. In: Sato R (ed) An experimental study of homing behavior of chum salmon. US-Japan cooperative Science Program

Valcana T (1979) The role of triiodothyroxine (T_3) receptors in brain development. In: Mersami E, Brazier MA (eds) Neural growth and differentiation. Raven Press, New York, pp 37–49

Van der Crack G, Eales JG (1980) Saturable 3, 5, 3'Triiodo-L-thyronine-binding sites in rat liver nuclei of rainbow trout (*Salmo gairdneri*, Richardson). Gen Comp Endocrinol 42:437–448

Van Overbeeke AP, McBride JR (1967) The pituitary gland of the sockeye (*Oncorhynchus nerka*) during sexual maturation and spawning. J Fish Res Board Can 24:1791–1810

Vreeland RR, Wahle RJ, Arp AH (1975) Homing behavior and contribution to Columbia River fisheries of marked coho salmon released at two locations. Fish Bull 73:717–725

Wagner HH (1970) The parr-smolt metamorphosis in steelhead trout as affected by photoperiod and temperature. Ph D Thes, Oreg State Univ, Corvallis

Wagner HH (1974a) Seawater adaptation independent of photoperiod in steelhead trout (*Salmo gairdneri*). Can J Zool 52:805–812

Wagner HH (1974b) Photoperiod and temperature regulation of smolting in steelhead trout (*Salmo gairdneri*). Can J Zool 52:219–234

Wagner HH, Conte FP, Fessler JL (1969) Development of osmotic and ionic regulation in two races of chinook salmon *Oncorhynchus tshawytscha*. Comp Biochem Physiol 29:325–341

Wahle RJ (1975) Observations on homing and fishery contribution of multiple releases of coho salmon. 1975 Symposium on Salmon Homing (summary notes) Natl Mar Fish Serv, Seattle, Washington

Walker JC (1967) Odor discrimination in relation to homing in Atlantic salmon. M S Thes, Univ New Brunswick, Fredericton

Weber D (1975) Homing of adult steelhead trout in the Snake River Basin following alteration of normal downstream migration by transportation. 1975 Symposium on Salmon Homing (summary notes). Natl Mar Fish Ser, Seattle, Washington

Weisbart M (1968) Osmotic and ionic regulation in embryos, alevins, and fry of the five species of Pacific salmon. Can J Zool 46:385–397

White HC (1934) Some facts and theories concerning Atlantic salmon. Trans Am Fish Soc 64:360–362

White HC (1936) The homing of salmon in the Apple River, N S J Biol Board Can 2:391–400

Whitehead C, Bromage NR, Forster JRM (1978) Seasonal changes in reproductive function of the rainbow trout (*Salmo gairdneri*). J Fish Biol 12:601–608

Wickett WP (1958) Adult returns of pink salmon from the 1954 Fraser River planting. Fish Res Board Can Prog Rep No 111:18–19

Winter J (1976) Radiotracking of rainbow trout in Lake Superior. Ph D Thes, Univ Minn, Minneapolis, 347 pp

Wisby WJ (1952) Olfactory responses of fishes as related to parent stream behavior. Ph D Thes, Univ Wis, Madison

Wisby WJ, Hasler AD (1954) The effect of olfactory occlusion on migrating silver salmon (*O. kisutch*). J Fish Res Board Can 11:472–478

Woodhead AD (1975) Endocrine physiology of fish migration. Oceanogr Mar Biol Annu Rev 13:287–382

Zambrano D, Nishioka RS, Bern HA (1972) The innervation of the pituitary gland of fishes. Its origin, nature and significance. In: Brain-endocrine interaction. Structure and function. Int Symp Munich 1971. Karger, Basel, pp 50–66

Zaugg WS, McLain LR (1969) Inorganic salt effects on growth, salt water adaptation, and gill ATPase of Pacific salmon. In: Neuhaus OW, Halver JE (eds) Fish in research. Academic Press, London New York, pp 295–306

Zaugg WS, McLain LR (1970) Adenosinetriphosphatase activity in gills of salmonids: seasonal variations and salt water influence in coho salmon, *Oncorhynchus kisutch*. Comp Biochem Physiol 35B:587–596

Zaugg WS, McLain LR (1971) Gill sampling as a method of following biochemical changes: ATPase activities altered by ouabain injection and salt water adaptation. Comp Biochem Physiol 38B:501–506

Zaugg WS, McLain LR (1972) Changes in gill adenosinetriphosphatase activity associated with parr-smolt transformation in steelhead trout, coho, and spring chinook salmon. J Fish Res Board Can 29:167–171

Zaugg WS, Wagner HH (1973) Gill ATPase activity related to parr-smolt transformation and migration in steelhead trout (*Salmo gairdneri*): influence of photoperiod and temperature. Comp Biochem Physiol 45B:955–965

Subject Index

Zoophysiology

formerly
Zoophysiology and Ecology
Coordination Editor: **D. S. Farner**
Editors: **D. S. Farner, B. Heinrich, K. Johansen, H. Langer, G. Neuweiler, D. J. Randall**

Volume 1
P. J. Bentley
Endocrines and Osmoregulation
A Comparative Account of the Regulation of Water and Salt in Vertebrates

1971. 29 figures. XVI, 300 pages
ISBN 3-540-05273-9

Volume 2
L. Irving
Arctic Life of Birds and Mammals
Including Man

1972. 59 figures. XI, 192 pages
ISBN 3-540-05801-X

Contents: Environment of Artic Life. - Mammals of the Arctic. - Arctic Land Birds and Their Migrations. - Maintenance of Arctic Populations: Birds. - Maintenance of Arctic Populations of Mammals. - Warm Temperature of Birds and Mammals. - Maintenance of Warmth by Variable Insulation. - Metabolic Supply of Heat. - Heterothermic Operation of Homeotherms. - Size and Seasonal Change in Dimensions. - Insulation of Man.

Volume 3
A. E. Needham
The Significance of Zoochromes
1974. 54 figures. XX, 429 pages
ISBN 3-540-06331-5

Contents: General. - The Nature and Distribution of Zoochromes. - Physiological Functions of Zoochromes. - Biochemical Functions of Zoochromes. - The Significance of Zoochromes for Reproduction and Development. - Evidence from Chromogenesis in the Individual. - Evolutionary Evidence and General Assessment.

Volume 4/5
A. C. Neville
Biology of the Arthropod Cuticle
1975. 233 figures. XVI, 448 pages
ISBN 3-540-07081-8

Contents: General Structure of Integument. - The Structural Macromolecules. - Molecular Cross-Linking. - Supermolecular Architecture. - Physiological Aspects. - Calcification. - Physical Properties. - Phylogenetical Aspects. - Outstanding Problems.

Volume 6
K. Schmidt-Koenig
Migration and Homing in Animals
1975. 64 figures, 2 tables. XII, 99 pages
ISBN 3-540-07433-3

From the contents: Field Performance in Orientation and Experimental and Theoretical Analysis in Crustaceans and Spiders, Locusts, Bees, Butterflies, Fishes, Amphibians, Reptiles, Birds, Mammals: Bats, Whales and Terrestrial Mammals. - Conclusion. - Appendix: Some Statistical Methods for the Analysis of Animal Orientation Data. - References. - Subject Index.

Volume 7
E. Curio
The Ethology of Predation
1976. 70 figures, 16 tables. X, 250 pages
ISBN 3-540-07720-0

Contents: Introduction. - Internal Factors. - Searching for Prey. - Prey Recognition. - Prey Selection. - Hunting for Prey.

Volume 8
W. Leuthold
African Ungulates
A Comparative Review of Their Ethology and Behavioral Ecology

1977. 55 figures, 7 tables. XIII, 307 pages
ISBN 3-540-07951-3

Contents: General Background. - Comparative Review of Nonsocial Behavior. - Social Behavior. - Social Organization. - Behavior and Ecological Adaptation.

Springer-Verlag
Berlin
Heidelberg
New York
Tokyo

Zoophysiology

formerly
Zoophysiology and Ecology
Coordination Editor: **D. S. Farner**
Editors: **D. S. Farner, B. Heinrich, K. Johansen, H. Langer, G. Neuweiler, D. J. Randall**

Volume 9
E. B. Edney
Water Balance in Land Arthropods
1977. 109 figures, 36 tables. XII, 282 pages
ISBN 3-540-08084-8

Contents: Introduction. – Water content. – Water loss – cuticular. – Water loss – respiratory. – Water loss by evaporative cooling. – Excretion and osmoregulation. – Uptake of liquid water. – Metabolic water. – Absorption of water vapour. – Water balance in eggs. – Conclusions.

Volume 10
H.-U. Thiele
Carabid Beetles in Their Environments
A Study on Habitat Selection by Adaptations in Physiology and Behaviour

Translated from the German by J. Wieser
1977. 152 figures, 58 tables. XVII, 369 pages
ISBN 3-540-08306-5

Contents: Variations in the Body Structure of Carabids in Adaptation to Environment and Mode of Life. – Quantitative Investigations on the Distribution of Carabids. – The Connections Between Carabids and Biotic Factors in the Ecosystem. – Man and the Ground Beetles. – The Differences in Distribution of Carabids in the Environment: Reactions to Abiotic Factors and Their Significance in Habitat Affinity. – Ecological Aspects of Activity Patterns in Carabids. – Choice of Habitat: The Influence of Connection Between Demands Upon Environmental Factors and Activity Rhythms. – Dispersal and Dispersal Power of Carabid Beetles. – Ecological Aspects of the Evolution of Carabids. – Concerning the Reasons Underlying Species Profusion Manifest by the Carabids.

Volume 11
M. H. A. Keenleyside
Diversity and Adaption in Fish Behaviour
1979. 67 figures, 15 tables. XIII, 208 pages
ISBN 3-540-09587-X

Contents: Locomotion. – Feeding Behaviour. – Anti-Predator Behaviour. – Selection and Preparation of Spawning Site. – Breeding Behaviour. – Parental Behaviour. – Social Organization. – References. – Systematic Index. – Subject Index.

Volume 12
E. Skadhauge
Osmoregulation in Birds
1981. 42 figures. X, 203 pages
ISBN 3-540-10546-8

Contents: Introduction. – Intake of Water and Sodium Chloride. – Uptake Through the Gut. – Evaporation. – Function of the Kidney. – Function of the Cloaca. – Function of the Salt Gland. – Interaction Among the Excretory Organs. – A Brief Survey of Hormones and Osmoregulation. – Problems of Life in the Desert, of Migration, and of Egg-Laying. – References. – Systematic and Species Index. – Subject Index.

Volume 13
S. Nilsson
Autonomic Nervre Function in the Vertebrates
1983. 83 figures. XIV, 253 pages
ISBN 3-540-12124-2

Contents: Introduction. – Anatomy of the Vertebrate Autonomic Nervous Systems. – Neurotransmission. – Receptors for Transmitter Substances. – Chemical Tools. – Chromaffin Tissue. – The Circulatory System. – Spleen. – The Alimentary Canal. – Swimbladder and Lung. – Urinary Bladder. – Iris. – Chromatophores. – Concluding Remarks. – References. – Subject Index.

Springer-Verlag
Berlin
Heidelberg
New York
Tokyo

GPSR Compliance
The European Union's (EU) General Product Safety Regulation (GPSR) is a set of rules that requires consumer products to be safe and our obligations to ensure this.

If you have any concerns about our products, you can contact us on

ProductSafety@springernature.com

In case Publisher is established outside the EU, the EU authorized representative is:

Springer Nature Customer Service Center GmbH
Europaplatz 3
69115 Heidelberg, Germany

www.ingramcontent.com/pod-product-compliance
Lightning Source LLC
Chambersburg PA
CBHW061216240926
51846CB00006B/1244
* 9 7 8 3 6 4 2 8 2 0 7 2 4 *